ACS Organic Chemistry Study Guide 2025-2026

Complete Review + 420 Questions and Detailed Answer Explanations for the American Chemical Society Exam (6 Full-Length Exams)

Table of Contents

****The digital flash cards can be downloaded via the QR Code provided at the end of this book. If it does not work for you, please contact info@newstonetestprep.com and mention the title of this book.**

Introduction

The American Chemical Society (ACS) Organic Chemistry Exam is a standardized assessment designed to evaluate a student's comprehensive knowledge of undergraduate organic chemistry. Recognized by chemistry departments across the United States, this exam is often used as a final assessment in organic chemistry courses and serves as a benchmark for academic achievement in the field.

Background

Organic chemistry is a core component of undergraduate science education. It provides the framework for us to understand the structure, properties, reactivity, and synthesis of carbon-based compounds. The exam assesses students' grasp of this subject. It covers a wide range of topics, each essential to the discipline and its real-world applications.

Key content areas tested on the exam include:

- Nomenclature: Accurate naming of organic molecules based on International Union of Pure and Applied Chemistry (IUPAC) conventions.
- Structure, Hybridization, Resonance, and Aromaticity: Foundational concepts that influence molecular behavior and reactivity.
- Acids and Bases: How acid-base properties affect molecular stability and transformations.
- Stereochemistry: Spatial arrangements, chirality, and isomerism.
- Substitution and Elimination Mechanisms: Detailed knowledge of SN1, SN2, E1, and E2 pathways.
- Addition Reactions: Mechanistic pathways for reactions that involve alkenes, alkynes, and related functional groups.
- Spectroscopy: Interpretation of IR, nuclear magnetic resonance (NMR), and mass spectra to deduce molecular structures.
- Radical Reactions: Mechanisms and applications of free-radical processes.
- Conjugated and Aromatic Systems: Behavior of delocalized systems and aromatic stability.
- Aromatic Reactions: Electrophilic and nucleophilic substitution mechanisms in aromatic compounds.
- Carbonyl and Enolate Chemistry: Key transformations that involve aldehydes, ketones, enols, and enolates.
- Oxidation and Reduction: Major organic redox reactions and reagent use.

- Synthesis and Multistep Pathways: Coherent synthetic strategies designed from multiple transformations.
- Applications of Organic Chemistry: Translate knowledge to biological, industrial, and pharmaceutical contexts.

Chapter 1: IUPAC Nomenclature in Organic Chemistry

The nomenclature of organic chemistry, as defined by IUPAC, is a universally accepted framework that assigns each organic compound a distinct name that accurately represents its molecular structure. IUPAC nomenclature eliminates the ambiguity in organic compound names. Each name corresponds to a specific structure, which allows chemists to communicate effectively. The IUPAC system is based on systematic rules that dictate how to construct names from the molecular structure of compounds. This includes the identification of the longest carbon chain, functional groups, and substituents. IUPAC nomenclature facilitates collaboration and knowledge sharing across different regions and disciplines because it provides a common language for chemists worldwide.

Core Structure of IUPAC Names

Given the extensive variety of isomers and potential structures in organic chemistry, it can be quite difficult to keep track of all these configurations. Therefore, a systematic approach to organic compound names is essential. The IUPAC has established the conventions for this nomenclature.

An IUPAC name consists of several components:

- **Stereochemistry Identifier:** This denotes the stereochemical aspects of the molecule.
- **Locant/Substituents:** This specifies the locations and names of any groups attached to the main chain.
- **Parent Chain:** The longest continuous carbon chain that includes the principal functional group and provides the lowest possible locants for substituents and functional groups, not necessarily the one allowing the maximum number of substituents.
- **Suffix:** This indicates the highest-priority functional group present in the compound.

The root word indicates the number of carbon atoms in the parent chain, while the prefix identifies any side chains or substituents. The suffix also conveys the type of principal functional group and the saturation level (single, double, or triple bonds). Numerical locants are used to precisely indicate the positions of substituents and functional groups along the parent chain.

Example:

Compound: 3-methylpent-2-ene.

- **Stereochemistry Identifier:** None specified in this name (no E/Z or R/S designations).
- **Locant/Substituents:** "3-methyl" indicates a methyl group (–CH_3) is attached to the third carbon of the main chain. The double bond starts at carbon 2.
- **Parent Chain:** "pent" signifies that the longest continuous carbon chain has five carbon atoms.
- **Suffix:** "2-ene" indicates that there is a double bond starting at the second carbon.

Thus, the name 3-methylpent-2-ene conveys that the compound has a five-carbon chain (pent) with a double bond between carbons 2 and 3, and a methyl group attached to carbon 3.

Fundamental Steps in IUPAC Nomenclature

The process by which an organic compound is named follows a systematic approach that consists of several essential steps. First, the principal functional group is identified, as this group carries the highest priority and determines the suffix and overall structure of the name. Next, the longest parent chain is selected. It must include the highest-priority functional group and as many multiple bonds as possible. Once the chain is established, it is numbered in a way that assigns the lowest possible locants to the principal functional group and any multiple bonds present.

After this, all branches and substituents need to be identified and named, with their positions indicated in alphabetical order as prefixes. The complete name is a combination of the prefixes, root word, and suffixes. Locants are then inserted where necessary to clarify positions.

Functional Group Prioritization

Functional groups are ranked based on their priority, which affects both the suffix used and the numbering of the parent chain. The standard order of importance for IUPAC naming is as follows:

- Carboxylic acids > sulfonic acids > esters > acid chlorides > amides > nitriles > aldehydes > ketones > alcohols > thiols > amines > alkenes > alkynes > ethers > alkanes > halides.

The highest-priority group dictates the suffix, while the lower-priority groups are designated as prefixes.

For example, let's consider the compound 4-hydroxy-3-methylpentanoic acid.

Identify Functional Groups

- Carboxylic Acid: The presence of the -oic acid suffix indicates that the highest-priority functional group is a carboxylic acid.
- Alcohol: The hydroxy- prefix indicates the presence of an alcohol functional group.

Based on IUPAC priority, carboxylic acids take precedence over alcohols. Thus, the carboxylic acid dictates the suffix of the name, which is -oic acid.

Number the Parent Chain

The longest carbon chain includes the carboxylic acid, which is assigned position 1. The alcohol group (hydroxy) is on carbon 2 in the compound 4-hydroxy-3-methylpentanoic acid.

Hydrocarbon Nomenclature

Alkanes

For alkanes, straight-chain compounds utilize the suffix -ane (e.g., hexane). Alkanes are saturated hydrocarbons, which means they cannot accommodate any additional hydrogen atoms in their structure. They follow the general formula $C_nH2_{n+}2$, where *n* is the number of carbon atoms (for acyclic alkanes). Alkanes are also known as aliphatic hydrocarbons or paraffins. It is important to note that the suffix "-ane" in their names indicates that they belong to the alkane class of compounds.

Branched alkanes require you to identify the longest chain and number it to provide the lowest locants for substituents, which should be listed alphabetically (e.g., 3-methylhexane).

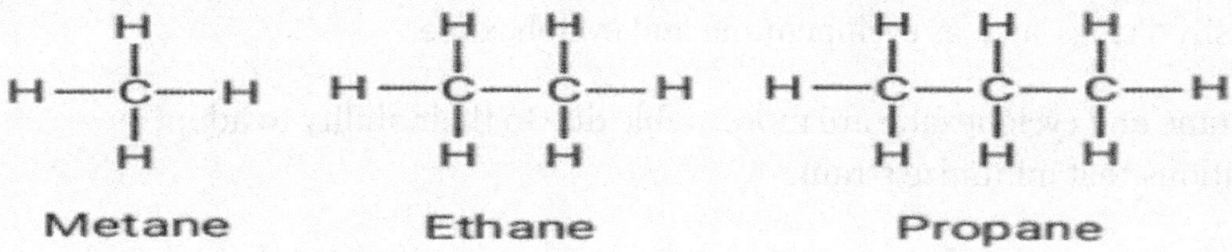

Metane Ethane Propane

Fig 1: Different examples of alkanes.

Cycloalkanes

Cycloalkanes are a class of saturated hydrocarbons characterized by ring structures. They follow the general formula C_nH_{2n}, which differs from acyclic alkanes due to ring closure. This formula distinguishes them from linear alkanes, which follow the formula C_nH_{2n+2}.

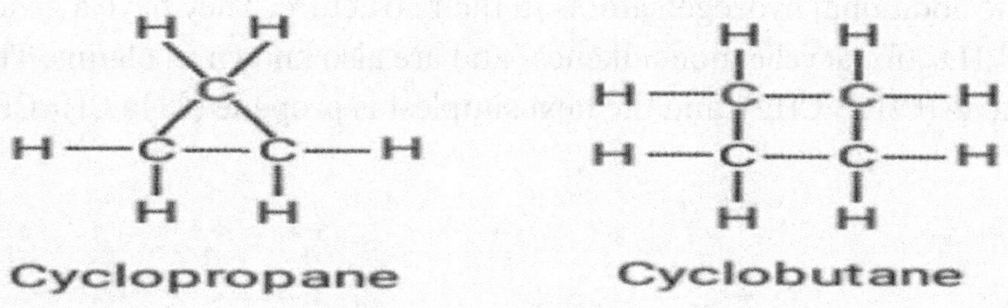

Fig 2: Different examples of cycloalkanes.

Cycloalkanes form closed loops or rings of carbon atoms. Each carbon atom in the ring is bonded to two other carbon atoms and has enough hydrogen atoms to fulfill its tetravalency (four bonds total). The simplest cycloalkanes include three to six carbon atoms, such as cyclopropane (C_3H_6), cyclobutane (C_4H_8), and cyclopentane (C_5H_{10}).

Cycloalkanes are saturated hydrocarbons and do not have any units of unsaturation; the presence of a ring does not count as unsaturation in IUPAC nomenclature. This means they are less saturated (fewer hydrogen atoms) compared to their straight-chain counterparts. Unsaturation can influence the chemical reactivity and physical properties of these compounds.

The stability of cycloalkanes varies with ring size. Smaller rings, like cyclopropane and cyclobutane, experience significant angle strain due to bond angles deviating substantially from the ideal tetrahedral angle of 109.5°, making them less stable than medium-sized rings such as cyclopentane and cyclohexane.

Cyclopentane and cyclohexane are more stable due to their ability to adopt conformations that minimize strain.

Cycloalkanes can exhibit various conformations. For instance, cyclohexane can adopt chair and boat forms, which greatly influence its stability and reactivity. Due to their cyclic nature, they have increased intermolecular forces. They generally have higher boiling points than their straight-chain counterparts. They are usually colorless and have low solubility in water but are soluble in organic solvents.

Alkenes

In the case of alkenes, the suffix -ene indicates the presence of double bonds, and the chain must be numbered to ensure the double bond receives the lowest possible number (e.g., hex-2-ene). Alkenes are unsaturated hydrocarbons, which means they can accommodate additional hydrogen atoms in their structure. They have a general formula of C_nH_{2n} (for acyclic monoalkenes) and are also known as olefins. The simplest alkene is ethene ($CH_2=CH_2$), and the next simplest is propene ($CH_3CH=CH_2$).

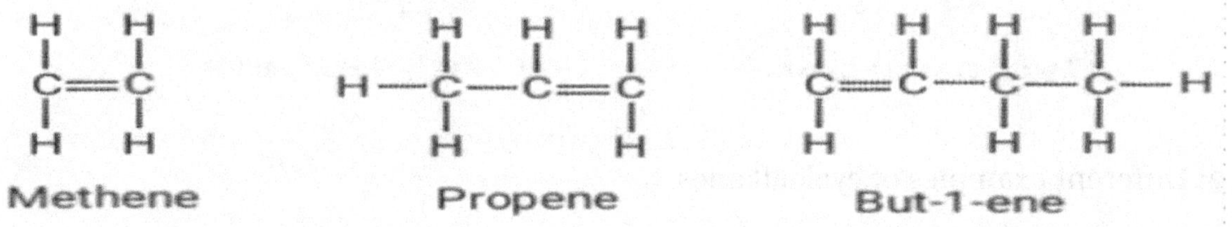

Methene Propene But-1-ene

Fig 3: Different examples of alkenes.

The presence of a double bond makes alkenes more reactive than alkanes. This enables them to participate in various chemical reactions, such as addition reactions. Alkenes can exhibit geometric (cis/trans) isomerism due to the restricted rotation around the double bond. Alkenes generally have slightly higher boiling points than alkanes of similar molecular weight due to increased polarizability from the double bond.

Alkynes

Similarly, alkynes use the suffix -yne for triple bonds, with numbering that gives the lowest locant to the triple bond (e.g., hex-3-yne). They are generally more reactive than alkenes due to the presence of a triple bond. This allows them to undergo various addition reactions. Alkynes often have boiling points comparable to or slightly higher than alkanes and alkenes of similar molecular weight, depending on molecular structure, due to linear geometry and π-bond interactions, not stronger intermolecular forces from the triple bond itself.

H—C≡C—H
Ethyne

H—C≡C—CH₃
Propyne

H—C≡C—CH₂—CH₃
1-Butyne

Fig 4: Different examples of alkynes.

When a compound contains both double and triple bonds, the naming convention combines the suffixes into “-enyne.” In such cases, priority is given to the bond that receives the lower number. For example, in pent-1-en-4-yne, the double bond is at position 1, while the triple bond is at position 4, but this name is incorrect because the chain must be numbered to give the first multiple bond the lowest possible number; the correct name is pent-4-en-1-yne.

Examples of Alkane, Alkene, and Alkyne Names

Straight-Chain Alkane

- Example: Hexane. A straight-chain alkane with six carbon atoms.

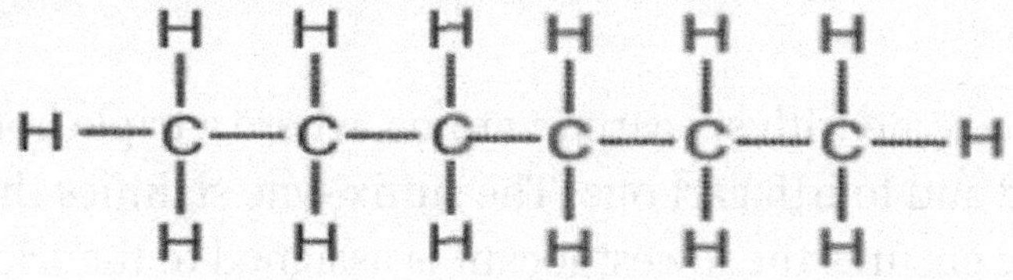

Fig 5: Structure of hexane.

Branched Alkane

- Example: 3-methylhexane. A branched alkane where the longest chain has six carbon atoms (hexane), with a methyl group attached to the third carbon.

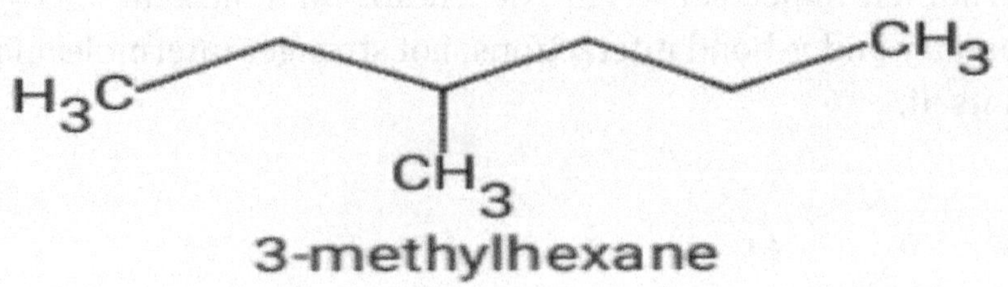

3-methylhexane

Fig 6: Structure of 3-methylhexane.

<u>Alkenes</u>

- Example: Hex-2-ene. An alkene with six carbon atoms, where a double bond starts at the second carbon and extends to the third. The suffix -ene indicates the presence of the double bond, and the chain is numbered to give the lowest locant to the double bond.

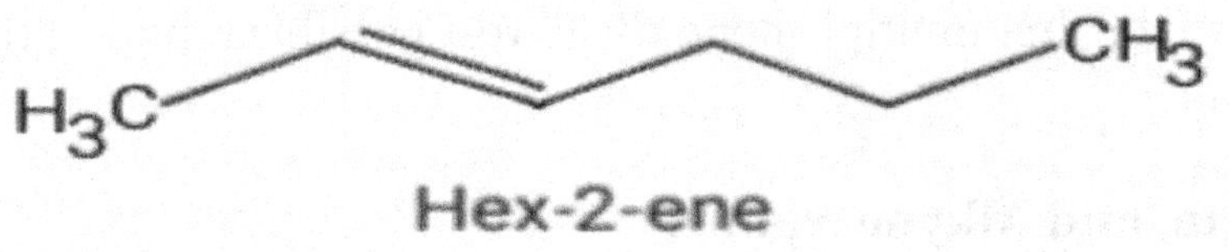

Hex-2-ene

Fig 7: Structure of hex-2-ene.

<u>Alkynes</u>

- Example: Hex-3-yne. An alkyne with six carbon atoms, where a triple bond is located between the third and fourth carbons. The suffix -yne signifies the triple bond, and the numbering ensures the lowest locant is assigned to the triple bond.

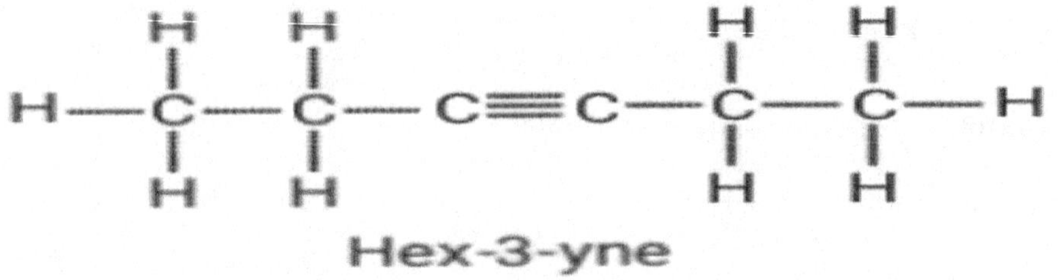

Hex-3-yne

Fig 8: Structure of hex-3-yne.

Compounds with Both Double and Triple Bonds

- Example: Pent-1-en-4-yne. A compound with five carbon atoms that contains both a double bond starting at the first carbon and a triple bond starting at the fourth carbon. The naming convention combines the suffixes to form -enyne and prioritizes the lower-numbered double bond.

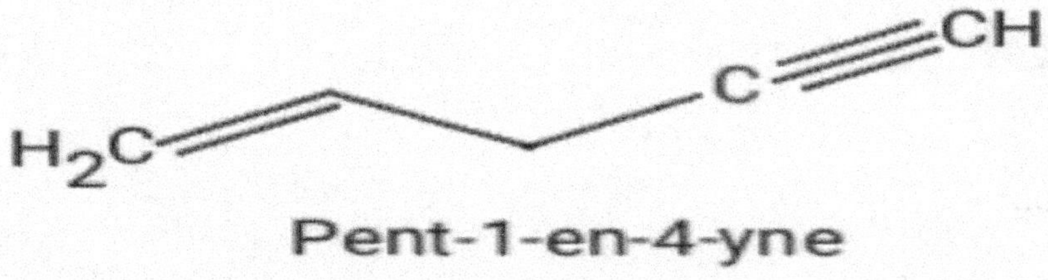

Fig 9: Structure of pent-1-en-4-yne.

Functional Group Naming

Different functional groups have specific suffixes and prefixes associated with them. For example, alcohols use the suffix -ol (as in butan-2-ol) and can be designated with the prefix hydroxy- when they are lower in priority. Ethers are named using alkoxy- (e.g., methoxy-). Aldehydes and ketones use -al (e.g., pentanal) and -one (e.g., pentan-2-one), respectively. Carboxylic acids are identified with the suffix -oic acid (e.g., butanoic acid) and use the prefix carboxy- only when the acid group is not the principal functional group. Amines use the suffix -amine and are prefixed with amino-, while nitriles are denoted with the suffix -nitrile and can use the prefix cyano-. When a functional group is repeated, multiplying affixes like di- or tri- are used, such as in -diol for two alcohol groups.

Complex and Cyclic Structures

For bicyclic compounds, the nomenclature follows the bicyclo [m.n.p] notation. m, n, and p denote the number of carbons in each bridge, with the total number of carbons calculated as $m + n + p + 2$ (e.g., bicyclo[2.2.1]heptane). Aromatic compounds, particularly those derived from benzene, are named using numerical locants or the ortho/meta/para system for disubstituted benzenes (e.g., 1,3-dichlorobenzene corresponds to meta-dichlorobenzene). Common names such as toluene and phenol are often retained for simplicity.

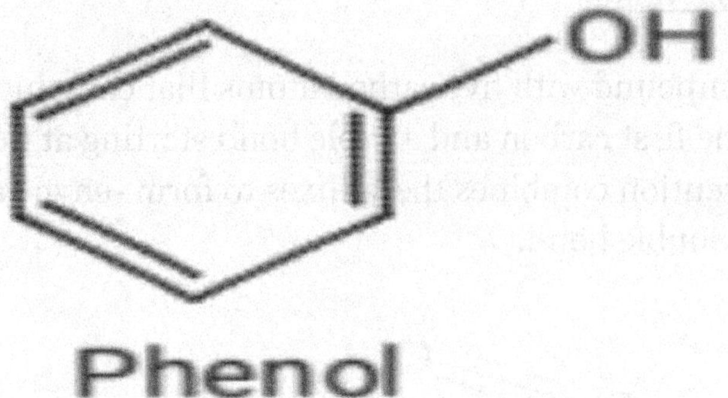

Fig 10: Structure of phenol.

Stereochemical Nomenclature

Geometric Isomerism

Cis/trans isomerism occurs due to restricted rotation around a double bond or within a cyclic structure. This allows substituents to be positioned differently in space but maintain the same connectivity. The terms "cis" and "trans" originate from Latin and mean "on this side" and "across," respectively. They refer to the relative positions of these substituents:

- Cis isomer: Substituents are located on the same side of the double bond or ring.
- Trans isomer: Substituents are positioned on opposite sides.

This isomerism is a type of configurational isomerism. If you want to change between isomers, this requires you to break bonds or overcome significant energy barriers, such as rotating around a double bond, which is generally not feasible under standard conditions.

Examples in Alkenes

Take but-2-ene, a straightforward alkene with two methyl groups attached to the carbons of the double bond:

- Cis-but-2-ene: Both methyl groups are on the same side of the double bond.
- Trans-but-2-ene: The methyl groups are on opposing sides.

These isomers exhibit different physical properties, such as boiling points and densities, due to their unique spatial arrangements.

Cis/Trans in Cyclic Compounds

In cyclic compounds, rotation around ring bonds is restricted. This locks substituents in specific positions relative to the ring.

- Cis: Both substituents are on the same face (either both above or below the plane).
- Trans: Substituents are located on opposite faces.

This naming system is commonly applied to cycloalkanes and other ring structures, where the orientation of substituents influences the molecule's shape and reactivity.

Limitations of Cis/Trans Nomenclature

The cis/trans nomenclature is effective when each carbon of the double bond has one identical substituent and allows clear definitions of "same" or "opposite" sides. However, it becomes inadequate in more complex scenarios:

- When substituents on the double-bond carbons differ.
- When there are more than two substituents, which results in ambiguity in "same" or "opposite" sides.

For instance, an alkene with four different groups attached to the double bond cannot be accurately described using cis or trans terminology and instead requires the E/Z nomenclature based on Cahn-Ingold-Prelog priority rules.

The E/Z Nomenclature System

To address the limitations of cis/trans nomenclature, the E/Z system was introduced as the IUPAC-approved method for naming alkene stereoisomers. It provides a clear way to describe the relative positions of substituents around a double bond, regardless of their complexity.

- E (entgegen): Means "opposite" in German.
- Z (zusammen): Means "together" in German.

Principle of Priority: Cahn-Ingold-Prelog Rules

The E/Z system utilizes the Cahn-Ingold-Prelog rules to determine the priority of substituents attached to each carbon of the double bond:

- The substituent with the higher atomic number attached directly to the double-bond carbon is given higher priority.
- In case of a tie, the next atoms in the chain are assessed until a difference is identified.
- Multiple bonds are treated as if the atoms are duplicated or triplicated.

Assign E or Z Configuration

Priorities are assigned to both carbons of the double bond.

- If the two highest-priority substituents are on the same side, the configuration is Z.
- If they are on opposite sides, the configuration is E.

Examples: For 2-butene, the cis isomer corresponds to the Z isomer only if the methyl groups are the highest-priority substituents on each carbon of the double bond, which they are in this case.

For a more complex alkene with substituents Br and I on one carbon and Cl and F on the other, priorities are assigned based on atomic number (I > Br and Cl > F). If the highest-priority groups (I and Cl) are on the same side, it is Z. If they are on opposite sides, it is E.

Advantages Over Cis/Trans

The E/Z system:

- Applies to alkenes with two, three, or four different substituents.
- Provides an absolute configuration rather than just a relative one.
- Eliminates ambiguity in how complex molecules are named.
- Applies to both cyclic and acyclic systems.

Practical Considerations of the Cis/Trans and E/Z Systems

When to Use Cis/Trans

- For simple alkenes with two identical substituents.
- For informal or quick descriptions.
- In cyclic compounds where substituents are clearly positioned on the same or opposite faces.

When to Use E/Z

- For alkenes with different substituents on the double-bond carbons.
- For formal nomenclature in publications and databases.
- When absolute stereochemical clarity is necessary.
- For molecules where cis/trans notation is ambiguous or unfeasible.

Common Mistakes and Tips

Several common mistakes can arise in IUPAC nomenclature. A frequent error is the failure to select the longest parent chain, which includes the highest-priority group. Numbering errors can occur when the lowest numbers are not assigned correctly to principal groups and multiple bonds. When substituents are alphabetized, it is important to disregard prefixes like di- or tri- and focus solely on the root name. For compounds with chiral centers or geometric isomerism, it is essential to include stereochemical descriptors (R/S, E/Z, cis/trans) as needed.

E does not always correspond to trans, nor does Z always correspond to cis; the E/Z system is based on atomic priority, whereas cis/trans relies on relative positioning, and the two may not always align. For example, a cis alkene can be E if the highest-priority groups are on opposite sides. Cis/trans descriptions can sometimes be represented by E/Z when priorities align, but not all E/Z cases can be described using cis/trans, making E/Z the more comprehensive system. The E/Z system is more comprehensive and should be preferred in academic and professional contexts.

Table 1: Key Suffixes and Prefixes

Functional Group	**Suffix**	**Prefix**	**Example Name**
Carboxylic acid	-oic acid	carboxy-	Butanoic acid
Ester	-oate	alkoxycarbonyl	Ethyl butanoate
Amide	-amide	carbamoyl-	Butanamide
Nitrile	-nitrile	cyano-	Pentanenitrile

Aldehyde	-al	formyl-	Pentanal
Ketone	-one	oxo-	Pentan-2-one
Alcohol	-ol	hydroxy-	Butan-2-ol
Amine	-amine	amino-	Butan-2-amine
Alkene	-ene	-	Hex-2-ene
Alkyne	-yne	-	Hex-3-yne

Chapter 2: Structure, Hybridization, Resonance, and Aromaticity

Bonding Theories: sp, sp^2, sp^3 Hybridization

Valence bond theory provides a framework for us to understand chemical bonding through the concept of orbital overlap. Hybridization is a key aspect of this theory, as it explains molecular geometry and bond angles by mixing atomic orbitals. Hybridization involves atomic orbitals that are mixed to create new, equivalent hybrid orbitals. These hybrid orbitals can then form sigma (σ) bonds, which are the primary bonds in most organic compounds. The three most common types of hybridization are sp, sp^2, and sp^3.

sp^3 Hybridization

In sp^3 hybridization, one s orbital combines with three p orbitals to form four equivalent sp^3 hybrid orbitals. This arrangement yields a tetrahedral geometry with bond angles of approximately 109.5°. Common examples include methane (CH_4), ethane (C_2H_6), and alcohols.

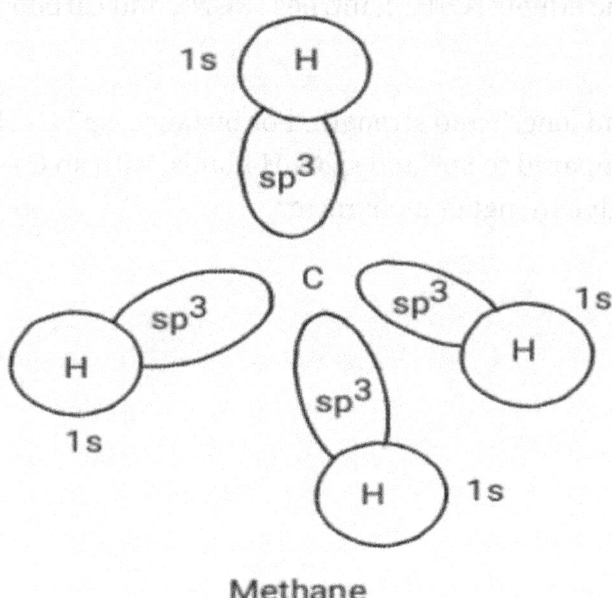

Fig 11: sp^3 hybridization.

sp^2 Hybridization

For sp^2 hybridization, one s orbital and two p orbitals combine to form three sp^2 hybrid orbitals, leaving one unhybridized p orbital. This results in a trigonal planar geometry with bond angles around 120°. Examples of sp^2 hybridization include ethene (C_2H_4), benzene (C_6H_6), and carbonyl compounds.

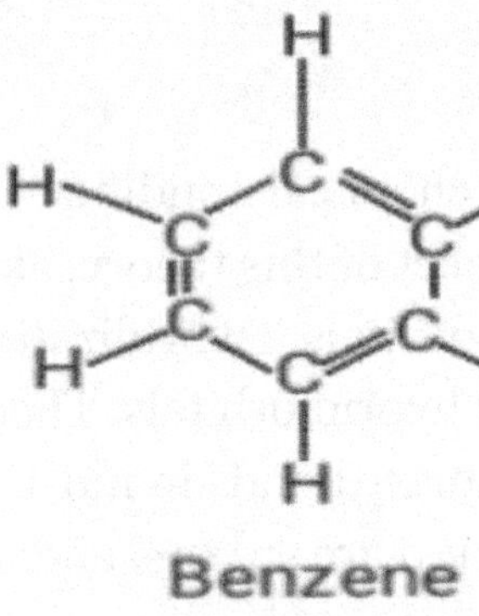

Benzene

Fig 12: sp^2 hybridization in benzene.

sp Hybridization

In sp hybridization, one s and one p orbital combine to form two sp hybrid orbitals, with two unhybridized p orbitals remaining. This configuration produces a linear geometry with bond angles of 180°. Examples include ethyne (C_2H_2), nitriles (RCN), and carbon dioxide (CO_2).

The degree of hybridization influences bond length and strength. For instance, sp^3 C–H bonds are generally longer and weaker compared to sp^2 and sp C–H bonds, with sp C–H bonds being the shortest and strongest due to higher s-character.

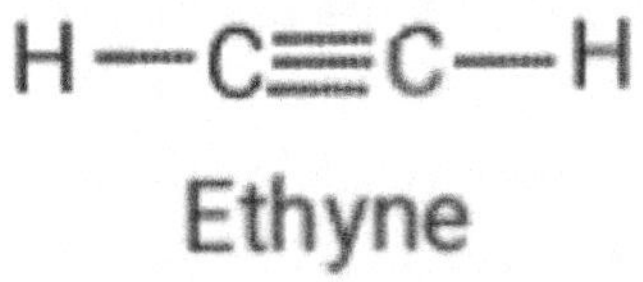

$H-C\equiv C-H$

Ethyne

Fig 13: sp hybridization in ethyne.

Resonance Structures

Resonance describes the delocalization of electrons in molecules with conjugated systems. It distributes electron density and stabilizes them. A single Lewis structure cannot adequately represent some compounds. Instead, multiple resonance structures are used to depict the actual electron distribution. Resonance structures are different Lewis structures for the same molecule that differ only in the placement of electrons.

The true structure is a hybrid of these resonance forms. Molecules with resonance are often more stable than those without. The delocalization of electrons lowers the overall energy of the molecule and makes it less reactive.

Rules for How to Draw Resonance Structures

- Only π electrons or lone pairs may move, while the positions of nuclei remain fixed.
- All resonance structures must comply with the octet rule, although there are exceptions for radicals or expanded octets.
- The most stable resonance structure will have minimal formal charges, negative charges on more electronegative atoms, and maximum bonding.

Examples

- **Ozone (O_3):** Exhibits two resonance structures in which the double bond and lone pairs are delocalized over the three oxygen atoms, resulting in equivalent bond lengths.

Fig 14: Resonance structures of ozone (O_3).

- **Carbonate Ion (CO_3^{2-}):** Has three equivalent resonance structures.
- **Benzene (C_6H_6):** Demonstrates resonance among structures with alternating double bonds, historically described as two Kekulé structures, but more accurately represented as a delocalized π electron cloud over all six carbon atoms.

Effects of Resonance in Organic Chemistry

Increased Stability

One of the main effects of resonance is the increased stability it provides to molecules. When resonance is present, a molecule can be represented by multiple Lewis structures, which show different arrangements of electrons. The true structure of the molecule is a hybrid of these resonance forms. This electron delocalization reduces the molecule's overall energy and makes it more stable than if represented by a single structure. For example, benzene, a classic aromatic chemical, has resonance. This adds to its extraordinary stability when compared to nonaromatic compounds.

Reduced Reactivity

Another significant consequence of resonance is the reduction in reactivity of certain molecules. Delocalized electrons are spread over several atoms rather than localized between two and can stabilize reactive intermediates. This stabilization makes some compounds less prone to undergo chemical reactions. For instance, in aromatic compounds, the delocalization of π electrons contributes to their resistance to electrophilic attack. This is a common reaction pathway for alkenes. This reduced reactivity is a key factor in the unique behavior of aromatic compounds in chemical reactions.

Influence on Acidity and Basicity

Resonance also plays an important role in the acidity and basicity of organic compounds. The presence of resonance can stabilize the negative charge that results when an acid donates a proton (H^+), which influences how readily a compound can act as an acid. For example, carboxylic acids exhibit resonance stabilization in their conjugate bases. This makes them stronger acids than alcohols. Similarly, basicity can be affected by resonance. For instance, when a base donates an electron pair, the resulting positive charge can be stabilized by resonance. This enhances the base's ability to accept protons.

Common Mistakes

It is important not to confuse resonance with equilibrium, as resonance structures do not represent the real states of the molecule but rather a hybrid. Additionally, it is important not to break σ bonds or alter atom positions when resonance structures are drawn.

Aromaticity

Aromatic compounds are a unique class of cyclic molecules that exhibit special stability due to resonance.

Criteria for Aromaticity (Hückel's Rule)

Aromatic compounds possess exceptional stability due to their cyclic, planar, conjugated structures.

Hückel's Rule

For a compound to be classified as aromatic, it must satisfy the following criteria:

- **Cyclic:** The molecule must contain a ring of atoms.
- **Planar:** All atoms in the ring, as well as any directly attached atoms, must be in the same plane to enable continuous overlap of the p orbitals in the conjugated system. This planarity is often achieved through the sp^2 hybridization of the ring atoms.
- **Conjugated:** The ring must feature a continuous arrangement of overlapping p orbitals. This is usually accomplished through alternating single and multiple bonds (double or triple bonds) or by including atoms with lone pairs of electrons that can engage in π bonding.
- **Hückel's Rule:** The total number of π electrons in the fully conjugated, planar, cyclic system must follow the formula $(4n + 2)$, where n is a non-negative integer (0, 1, 2, ...).

Molecules that satisfy all these criteria are considered aromatic and exhibit enhanced stability. They undergo electrophilic substitution reactions rather than addition reactions (characteristic of alkenes) and often show unique spectroscopic properties.

Examples:

- **Benzene:** Contains 6 π electrons: $n = 1$, which satisfies $4(1) + 2 = 6$.
- **Pyridine:** A heterocyclic aromatic compound that also contains 6 π electrons.
- **Naphthalene:** Has 10 π electrons: $n = 2$, which satisfies $4(2) + 2 = 10$.

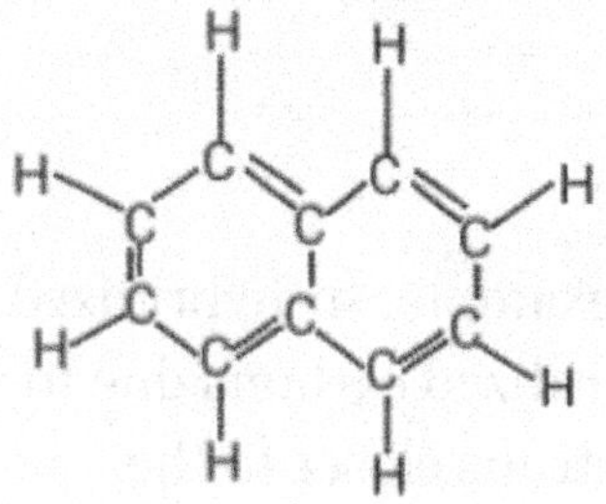

Fig 15: Aromatic compound.

Nonaromatic Compounds

Compounds that lack one or more of the aromaticity criteria are considered nonaromatic. For example, cyclooctatetraene is nonaromatic due to its nonplanar geometry and the presence of 8 π electrons.

Anti-Aromatic Compounds

These compounds meet all the criteria for aromaticity except that they do not contain 4n π electrons.

- Example: Cyclobutadiene has 4 π electrons (4n, n=1), is planar and conjugated, and is classified as anti-aromatic due to its instability.

Nonaromatic and Anti-Aromatic Examples

- **Nonaromatic:** Cycloheptatriene is not fully conjugated, and cyclopropane lacks π electrons.
- **Anti-Aromatic:** Cyclobutadiene, with 4 π electrons, and pentalene, with 8 π electrons, are both planar and conjugated but unstable due to their classification as anti-aromatic compounds.

Anti-aromatic compounds experience destabilization due to increased energy from electron delocalization.

Structure-Property Relationships

The structure of a molecule, both electronically and geometrically, significantly affects its physical and chemical properties.

Bond Length and Strength

Hybridization plays a key role in bond length and strength. For example, sp-hybridized carbons create shorter and stronger bonds than sp^2 and sp^3 hybridized carbons due to the greater s-character of their hybrid orbitals, which brings electrons closer to the nucleus. Triple bonds (one σ and two π) are shorter and stronger than double bonds (one σ and one π), which are also shorter and stronger than single bonds (one σ).

Molecular Shape and Polarity

Hybridization influences the geometry around an atom and affects the overall shape of the molecule. This shape and the polarity of individual bonds determine the molecule's overall polarity. Polar molecules have stronger intermolecular forces. This results in higher boiling points and different solubility characteristics compared to nonpolar molecules.

Reactivity

The presence of π bonds and the potential for resonance delocalization greatly impact a molecule's reactivity. Molecules with π bonds, such as alkenes and alkynes, typically undergo addition reactions, while aromatic compounds stabilized by resonance tend to undergo substitution rather than addition to preserve aromaticity. Aromatic compounds, known for their remarkable stability from cyclic delocalization of (4n+2) π electrons, exhibit distinct reactivity patterns.

Spectroscopic Properties

A molecule's electronic structure, shaped by hybridization and conjugation, determines its interaction with electromagnetic radiation. For instance, conjugated systems and aromatic compounds usually absorb light in the UV-Vis range. This produces characteristic spectra. Additionally, NMR spectroscopy is particularly sensitive to the electronic environment of various nuclei, influenced by bonding and resonance.

Inductive and Mesomeric Effects in Organic Chemistry

Inductive Effect

The inductive effect is a key concept in organic chemistry that refers to the permanent shift in electron density along a molecular chain. This shift arises from differences in electronegativity among the atoms or groups in the molecule. It mainly occurs through sigma (σ) bonds. This results in polarized bonds with partial positive (δ^+) and partial negative (δ^-) charges distributed along the molecular structure.

The inductive effect originates from the unequal sharing of electrons in a bond. When two atoms bond, the atom with greater electronegativity attracts the bonding electrons more strongly, which results in a displacement of electron density. For example, in a compound with carbon and chlorine, chlorine's higher electronegativity causes it to pull electron density toward itself. This results in a partial negative charge on chlorine and a partial positive charge on carbon.

Types of Inductive Effects

Inductive effects can be divided into two main types based on the substituents:

Electron-Withdrawing Inductive Effect (–I effect)

This effect occurs when groups or atoms more electronegative than carbon, like halogens or nitro groups, withdraw electron density through sigma bonds. This withdrawal stabilizes positive charges or destabilizes negative charges and affects the molecule's overall charge distribution. For instance, a nitro group ($-NO_2$) can attract electron density from adjacent carbon atoms, which stabilizes any existing positive charges.

Electron-Donating Inductive Effect (+I effect)

In contrast, alkyl groups and other electron-rich substituents donate electron density through sigma bonds. This donation enhances electron density on neighboring atoms, which can be vital for stabilizing negative charges. For example, a methyl group ($-CH_3$) adds electron density. This reduces a compound's acidity because it destabilizes its conjugate base.

Mechanism and Bond Polarization

The mechanism behind the inductive effect is based on bond polarization. In a covalent bond between atoms with different electronegativities, electrons are not shared equally. They shift toward the more electronegative atom, which creates bond polarization.

A usual example is water (H_2O). The oxygen atom, which is more electronegative than hydrogen, attracts the bonding electrons more effectively. This results in a partial negative charge (δ^-) on oxygen and a partial positive charge (δ^+) on hydrogen. This polarization helps us understand the inductive effect as it influences various molecular properties, including reactivity and dipole moments.

Influence on Molecular Properties

The inductive effect significantly impacts the acidity and basicity of compounds. Electron-withdrawing groups (EWGs) enhance a molecule's acidity because they stabilize the conjugate base through the –I effect. This lowers its energy and favors proton donation. Conversely, electron-donating groups (EDGs) reduce acidity through the +I effect. This increases electron density and destabilizes the conjugate base.

Furthermore, the inductive effect influences bond strengths, reaction pathways, and overall molecular stability. For instance, strong EWGs can increase the acidity of nearby C-H bonds (such as in α-hydrogens) by stabilizing the resulting conjugate base, making proton abstraction more favorable rather than causing bond weakening directly.

Mesomeric Effect (Resonance Effect)

The mesomeric effect, or resonance effect, involves the delocalization or movement of π electrons within a molecule through resonance structures. Unlike the inductive effect, which works through sigma bonds, the mesomeric effect is concerned with the movement of electrons in π bonds or between π bonds and adjacent lone pairs. This delocalization results in various resonance forms that collectively convey the true electronic structure of the molecule as a resonance hybrid with delocalized electron density.

Types of Mesomeric Effects

Mesomeric effects can also be classified based on the substituents involved:

Positive Mesomeric Effect (+M)

This effect occurs when EDGs inject electron density into the π system, which increases electron density and stabilizes the molecule. Groups that exhibit a positive mesomeric effect include those with adjacent lone pairs to π bonds, such as hydroxyl (–OH), alkoxy (–OR), and amino ($-NH_2$). These groups enhance the resonance stability of the π system and stabilize the molecular structure.

Negative Mesomeric Effect (–M)

On the other hand, EWGs pull electron density away from the π system, which decreases electron density and often destabilizes certain resonance forms. Groups like nitro ($-NO_2$), carbonyl (–CHO), and cyano (–CN) show a negative mesomeric effect. They withdraw electron density from the π system, which can stabilize certain resonance structures (e.g., in conjugate bases or electrophilic centers), but may reduce electron density in nucleophilic contexts. This affects the molecule's reactivity and stability.

Mechanism and Resonance Structures

The mesomeric effect results from the overlap of p orbitals and the movement of π electrons or lone pairs, which results in resonance structures. These structures vary only in the position of electrons, not in the arrangement of atoms. Even though these forms are theoretical, they provide valuable insights into the electronic structure of the molecule, represented as a hybrid of these resonance forms.

For example, in benzaldehyde, the aldehyde group pulls electron density (–M effect) and affects the aromatic ring's reactivity. In contrast, phenol exhibits a positive mesomeric effect. The hydroxyl group donates electron density (+M effect) to the aromatic system and enhances the molecule's stability.

Applications and Implications

Knowledge of both inductive and mesomeric effects is vital to help us predict organic compounds' behavior in chemical reactions. These electronic effects can greatly influence reaction mechanisms, stability, and the acidity or basicity of functional groups.

For example, in electrophilic aromatic substitution reactions, the presence of EDGs (through the +M effect) can increase the aromatic ring's reactivity and make it more prone to electrophilic attack. Conversely, EWGs can deactivate the ring and reduce its reactivity.

Both inductive and mesomeric effects are also vital in drug design and molecular biology. When chemists manipulate these electronic effects, they can develop molecules with specific properties. This enhances their effectiveness as pharmaceuticals or helps them target particular biological pathways.

Common Mistakes and Tips

- Hybridization can be misidentified, such as when one incorrectly assumes some atoms in benzene are not sp^2; in fact, all carbon atoms in benzene are sp^2 hybridized.
- Conjugation can be overlooked in resonance structures, and valid resonance forms can be missed.
- Hückel's rule can be misapplied if one neglects to consider planarity or the necessity of a fully conjugated π system.

Chapter 3: Acids and Bases in Organic Chemistry

Acids and bases play a pivotal role in organic reaction mechanisms and affect everything from reaction rates to product selectivity.

Arrhenius Definition

Arrhenius introduced one of the first definitions of acids and bases in 1884. He stated that acids are substances that produce H^+ ions in water, while bases produce OH^- ions. However, this definition needs to be revised to reflect current knowledge. In aqueous solutions, H^+ does not exist freely but associates with water molecules to form species such as hydronium (H_3O^+), the Zundel cation ($H_5O_2^+$), and the Eigen cation ($H_9O_4^+$), which are dynamic and interconvert in solution. Usually, only the hydronium ion is represented. Although the Arrhenius definition is helpful for water-based scenarios, it is limited and does not encompass the entire range of acid-base chemistry.

Brønsted–Lowry vs. Lewis Definitions

Brønsted–Lowry Acids and Bases

In the Brønsted–Lowry acid-base theory established in 1923, an acid is defined as a molecule that donates a proton (H^+), while a base is one that accepts a proton. This theory emphasizes the relationship between acids and bases: When an acid donates a proton to a base, it forms a conjugate base. The base becomes a conjugate acid when it accepts the proton.

Example:

- $HCl + H_2O \rightleftharpoons Cl^- + H_3O^+$

Fig 16: An HCl (Brønsted–Lowry acid) and H_2O (Brønsted–Lowry base) reaction produces a conjugate base (Cl^-) and acid (H_3O^+).

In this reaction, HCl donates a proton to water and produces hydronium (H_3O^+). Water acts as the base and accepts the proton from HCl. This results in the formation of chloride (Cl^-, the conjugate base of HCl) and hydronium (H_3O^+, the conjugate acid of H_2O).

Lewis Definitions of Acids and Bases

The Lewis definition expands our knowledge of acids and bases beyond the traditional Brønsted–Lowry (focused on proton transfer) and Arrhenius (which involves H^+/OH^- in water) definitions. According to G.N. Lewis, a Lewis acid is a species that can accept an electron pair, which makes it an electron pair acceptor. These acids are often electron-deficient and feature an incomplete octet or a positive charge. Examples include BF_3, $AlCl_3$, $FeCl_3$, and H^+.

A Lewis base is a species that donates an electron pair and acts as an electron pair donor. Lewis bases usually possess lone pairs of electrons. Examples include NH_3, H_2O, OH^-, and species with π bonds that can donate electrons.

The interaction between a Lewis acid and a Lewis base forms a coordinate covalent bond in which both electrons come from the Lewis base. The resulting species is known as an adduct or complex. For example, when ammonia (NH_3) reacts with boron trifluoride (BF_3), the nitrogen in ammonia donates its lone pair to the electron-deficient boron atom:

- $N..H_3 + BF_3 \longrightarrow H_3N{-}BF_3$

In this reaction, BF_3 acts as a Lewis acid, while NH_3 serves as a Lewis base.

Remember that Brønsted–Lowry acids are also Lewis acids (since they donate H^+, which is an electron pair acceptor). Brønsted–Lowry bases are also Lewis bases (as they accept H^+ since they donate an electron pair). However, the Lewis definition encompasses a broader range of acidic and basic species that do not necessarily involve proton transfer.

pKa and Strength of Acids/Bases

The acid dissociation constant (K_a) quantitatively measures the strength of an acid in solution. This indicates how well an acid (HA) dissociates into its conjugate base (A^-) and a proton (H^+) in water:

- $HA(aq) + H_2O(l) \rightleftharpoons H_3O^+(aq) + A^-(aq)$

The equilibrium constant for this reaction is:

- $K_a = [H3O+][A-]/[HA]$

A larger K_a value signifies a stronger acid, which indicates greater dissociation in solution and a higher concentration of H_3O^+.

For convenience, acidity is often expressed using the pK_a value, which is the negative logarithm (base 10) of K_a:

- $pK_a = -\log_{10}(K_a)$

The relationship between K_a and pK_a is inverse. Stronger acids have larger K_a values and smaller pK_a values (closer to or below zero), though pK_a values are typically positive for most acids. Weaker acids have smaller K_a values and larger pK_a values.

Similarly, the strength of a base is quantified by its base dissociation constant (K_β). This describes the equilibrium of a base (B) that reacts with water to form its conjugate acid (BH^+) and hydroxide ions (OH^-):

- $B(aq) + H_2O(l) \rightleftharpoons BH^+(aq) + OH^-(aq)$

The equilibrium constant is:

- $K_\beta = [BH+][OH-]/[B]$

A larger K_β indicates a stronger base. Just like pK_a, we can define pK_β:

- $pK_\beta = -\log_{10}(K_\beta)$

For a conjugate acid-base pair in aqueous solution, there is a useful relationship between their pK_a and pK_β values:

- $pK_a + pK_\beta = 14$ (at 25°C)

This relationship allows us to infer the strength of a base from the pK_a of its conjugate acid and vice versa. A strong acid corresponds to a weak conjugate base, while a weak acid correlates with a strong conjugate base.

Factors that Influence Acidity/Basicity

Several factors impact the stability of the conjugate base (in terms of acidity) or the ability of a base to accept a proton (in terms of basicity), which affects the overall acidity or basicity of a compound. Some of the factors are listed below.

Electronegativity

For atoms within the same period of the periodic table, acidity generally increases with higher electronegativity. A more electronegative atom can better stabilize the negative charge of the conjugate base, and this enhances acid strength. For example, $NH_3 < H_2O < HF$.

Conversely, basicity tends to decrease within the same period as electronegativity increases. A more electronegative atom holds its lone pair more tightly and makes it less likely to donate it to a proton.

Atomic Size (and Polarizability)

For atoms in the same group of the periodic table, acidity usually increases with larger atomic size. The negative charge of the conjugate base is spread over a greater volume, which causes increased stability. Larger atoms are also more polarizable, and this helps stabilize the conjugate base through solvation. For instance, $HF < HCl < HBr < HI$.

For the same group, basicity usually decreases as atomic size increases. Larger atoms have more diffuse lone pairs, which makes them less available to accept a proton.

Resonance

If the conjugate base can be stabilized by resonance, the acidity of the corresponding acid increases. Delocalization of the negative charge through resonance spreads the charge over multiple atoms. This enhances the stability of the conjugate base and increases acid strength. Carboxylic acids are more acidic than alcohols partly due to resonance stabilization of the carboxylate anion, which delocalizes the negative charge over two electronegative oxygen atoms, unlike alkoxide ions from alcohols that lack such delocalization.

Resonance can reduce basicity if the lone pair on the basic atom participates in resonance, which makes it less available for proton acceptance. For example, nitrogen in amides is significantly less basic than in amines because the lone pair on the amide nitrogen is delocalized through resonance with the carbonyl group.

Inductive Effects

EWGs near the acidic proton stabilize the conjugate base because they pull electron density away from the negative charge, which increases acidity. Electronegative atoms (like halogens, oxygen, and nitrogen) and groups with multiple electronegative atoms (e.g., NO_2, CN) serve as EWGs. For example, chloroacetic acid is more acidic than acetic acid due to the electron-withdrawing effect of chlorine.

EDGs near a basic site increase electron density on the basic atom. This enhances its ability to accept a proton and thus increases basicity. Alkyl groups are weak EDGs. For instance, alkylamines are generally more basic than ammonia. EWGs decrease basicity because they pull electron density away from the basic site.

Hybridization

The hybridization of the atom that holds the lone pair in the conjugate base (or the atom losing the proton) affects acidity. Higher s-character in the hybrid orbital means the electrons are held closer to the nucleus, which stabilizes the conjugate base. Therefore, the acidity of C-H bonds increases in the order $sp^3 < sp^2 < sp$. For example, terminal alkynes are more acidic than alkenes, which are more acidic than alkanes.

Similarly, the hybridization of the atom with the lone pair in the base influences basicity. A lone pair in an orbital with more s-character is more tightly held and less available for protonation, which results in lower basicity. For example, the nitrogen in an sp-hybridized nitrile is less basic than the nitrogen in an sp^2-hybridized imine, which is in turn less basic than the nitrogen in an sp^3-hybridized amine.

Applications to Reaction Mechanisms

Acids and bases play several key roles in chemical reactions.

- **Catalysis:** Both Brønsted and Lewis acids can act as catalysts by protonating a reactant to enhance its electrophilicity (making it more susceptible to nucleophilic attack) or by stabilizing a leaving group. Bases can catalyze reactions by deprotonating a reactant to generate a stronger nucleophile or by activating a leaving group.
- **Acid Catalysis:** In the acid-catalyzed hydration of an alkene, the acid (H_3O^+) protonates the alkene and forms a more electrophilic carbocation intermediate that is subsequently attacked by water.
- **Base Catalysis:** In aldol condensation under basic conditions, a base (e.g., OH^-) deprotonates an α-hydrogen of a carbonyl compound, which creates an enolate ion, a strong nucleophile.
- **Activation of Leaving Groups:** Acids can protonate leaving groups (e.g., converting OH^- to H_2O), which makes them better leaving groups since the departure of a neutral molecule is energetically more favorable than that of a negatively charged ion.

- **Generation of Nucleophiles and Electrophiles:** Bases can deprotonate weak acids to produce nucleophiles (e.g., the formation of alkoxides from alcohols). Lewis acids can coordinate with a lone pair of electrons to activate electrophiles, which increases the positive character of the electrophilic center.

$$CH_3CH_2OH + Na \longrightarrow CH_3CH_2O^-Na^+ + \frac{1}{2}H_2(g)$$

Ethanol   Sodium metal   Sodium ethoxide

- **Stabilization of Intermediates:** Resonance and inductive effects, which influence acidity and basicity, help stabilize charged intermediates (carbocations, carbanions) formed during reactions. The stability of these intermediates often dictates the reaction pathway and rate.

Chapter 4: Stereochemistry

Stereochemistry is a fundamental branch of chemistry that focuses on the three-dimensional arrangement of atoms in molecules and how this spatial orientation influences their chemical and physical properties. An important aspect of stereochemistry is the concept of isomers, which are compounds that share the same molecular formula but differ in structure or arrangement.

Isomer Types: Constitutional vs. Stereoisomers

Isomers are molecules with the same molecular formula but different arrangements of atoms. They can be broadly classified into two main types.

Constitutional Isomers (Structural Isomers)

Constitutional isomers have the same molecular formula but differ in how their atoms are connected. This variation in connectivity can lead to different functional groups, carbon skeletons, or positions of substituents.

Constitutional isomers can be subdivided into three main types: skeletal (chain) isomers, positional isomers, and functional isomers.

1) Skeletal (Chain) Isomers

Skeletal isomers differ in the branching of their carbon skeletons. These variations can significantly affect the properties of the compounds. For example, consider the two isomers of pentane:

- **n-Pentane:** This isomer has a straight-chain structure with five carbon atoms connected linearly (C_5H_{12}).

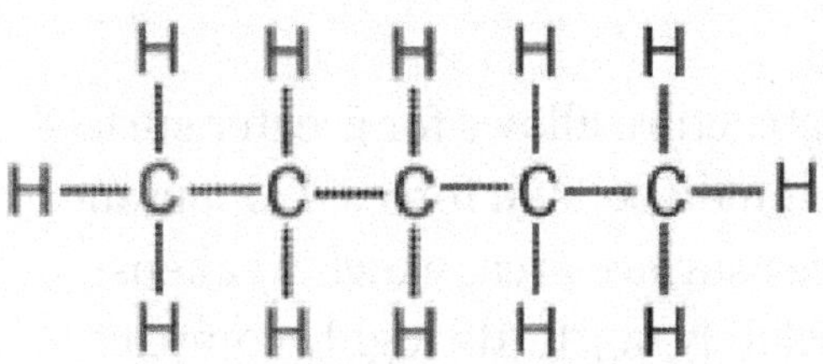

Fig 17: Structure of n-Pentane.

- **Isopentane (2-Methylbutane):** This branched isomer has a four-carbon chain with a methyl group on the second carbon, resulting in a different structural arrangement with the same molecular formula (C_5H_{12}).

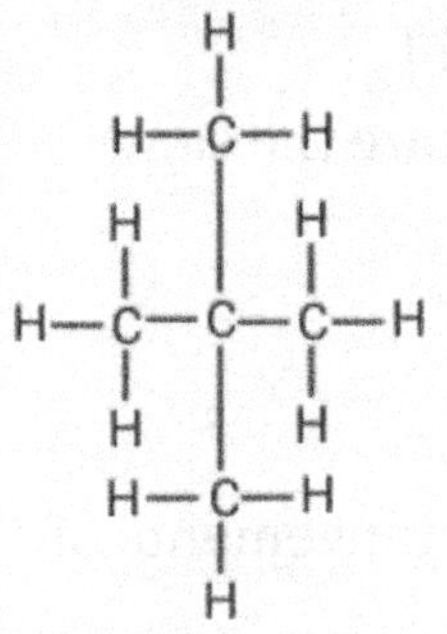

Fig 18: Structure of isopentane.

- **Neopentane (2,2-Dimethylpropane):** A central carbon atom bonded to three methyl groups and one hydrogen—$C(CH_3)_4$.

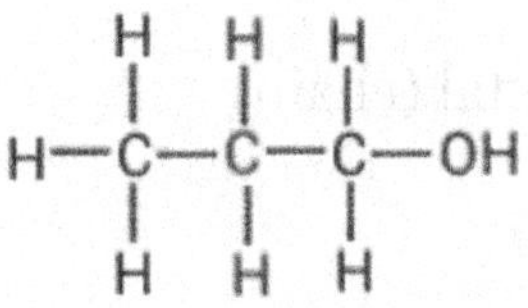

Fig 19: Structure of neopentane.

The difference in branching yields distinct boiling points and densities, which demonstrates how skeletal isomers can exhibit varied physical properties even if they have the same molecular formula. Boiling points are largely determined by the strength of intermolecular forces, which depend on molecular shape. Linear molecules have stronger van der Waals forces due to increased surface area, while branched molecules are more compact, reducing these forces.

n-Pentane boils at approximately 36.1°C. Its linear structure allows for greater surface contact between molecules, which strengthens London dispersion forces. Isopentane boils at approximately 27.8°C. The branching reduces surface area, which weakens intermolecular forces. Neopentane boils at approximately 9.5°C. Its highly compact, spherical structure minimizes surface contact between molecules, which reduces London dispersion forces and results in the lowest boiling point among the C_5H_{12} isomers.

2) Positional Isomers

Positional isomers vary in the location of functional groups within the carbon chain. Unlike skeletal isomers, which involve differences in the carbon backbone, positional isomers maintain the same carbon framework but vary in where the functional group or substituent is located. This type of isomerism is common in organic compounds, particularly in alcohols, alkenes, alkynes, and substituted aromatic compounds. This difference can influence the chemical reactivity and properties of the compounds. A notable example includes:

- **1-Propanol:** In this isomer, the hydroxyl group (OH) is attached to the terminal carbon atom (C_3H_8O).

```
   H   H   H
   |   |   |
H—C—C—C—OH
   |   |   |
   H   H   H
```

Fig 20: Structure of 1-propanol.

- **2-Propanol:** Here, the hydroxyl group is bonded to the middle carbon atom, which results in a different spatial arrangement (C_3H_8O).

```
   H  OH  H
   |   |   |
H—C—C—C—H
   |   |   |
   H   H   H
```

Fig 21: Structure of 2-propanol.

The positioning of the functional group alters the physical properties and reactivity of these alcohols. This illustrates the importance of functional group placement in determining compound behavior.

3) Functional Isomers

Functional isomers contain different functional groups, which can lead to distinct chemical properties and reactivity profiles. This type of isomerism is distinct from skeletal or positional isomerism, as it involves a fundamental change in the type of chemical bonding or functional group rather than the arrangement of the carbon skeleton or the position of a functional group. Functional isomers usually exhibit significantly different chemical reactivities because the functional group dictates how the molecule interacts with other substances.

For example:

- **Ethanol (C_2H_5OH):** This compound is an alcohol with a hydroxyl functional group; its molecular formula can also be written as C_2H_6O to represent its isomeric relationship with dimethyl ether.

```
   H  H
   |  |
H—C—C—O—H
   |  |
   H  H
```

Fig 22: Structure of ethanol.

- **Dimethyl Ether (CH_3OCH_3):** In contrast, this compound is an ether characterized by an oxygen atom bonded to two methyl groups.

```
   H     H
   |     |
H—C—O—C—H
   |     |
   H     H
```

Fig 23: Structure of dimethyl ether.

The presence of different functional groups not only affects the chemical reactivity but also the physical properties, such as boiling points and solubility of these compounds.

Stereoisomers

Stereoisomers are molecules that have the same molecular formula (same number and types of atoms) and connectivity (same sequence of bonded atoms) but differ in the three-dimensional arrangement of their atoms in space. This spatial difference causes distinct chemical and physical properties despite their identical composition. Stereoisomers are vital in fields like organic chemistry, biochemistry, and pharmacology. Their spatial arrangements can drastically affect their behavior, such as how they interact with biological systems. Stereoisomers are broadly classified into two main categories—enantiomers and diastereomers—based on their structural relationships.

Enantiomers

Enantiomers are a specific type of stereoisomer that are non-superimposable mirror images of each other, much like a left hand and a right hand. This property arises in chiral molecules, which lack symmetry elements such as a plane of symmetry, center of symmetry, or axis of symmetry. Chirality is often associated with a chiral center (or stereocenter), usually a carbon atom bonded to four different substituents. For example, in lactic acid—$CH_3CH(OH)COOH$—the central carbon bearing the hydroxyl group is a chiral center because it is bonded to four different groups: -H, -OH, $-CH_3$, and -COOH.

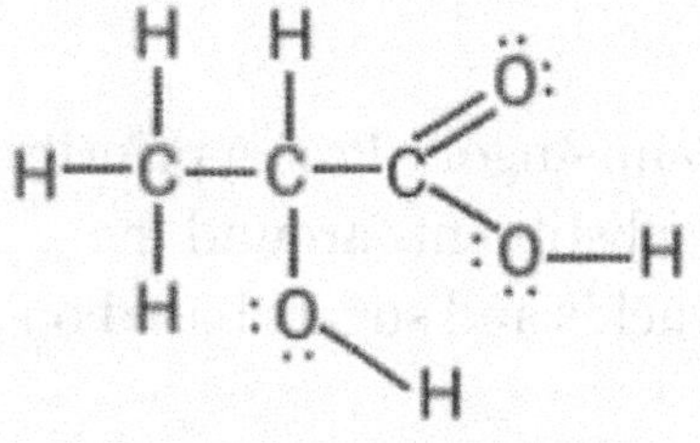

Fig 24: Structure of lactic acid.

<u>Characteristics of Enantiomers</u>

- **Mirror Image Relationship**: Enantiomers are exact reflections of each other. If you were to hold one enantiomer up to a mirror, its reflection would match the structure of the other enantiomer.

- **Identical Physical Properties**: Enantiomers have identical physical properties—melting points, boiling points, and most spectroscopic data (e.g., NMR)—but differ in their interaction with plane-polarized light and in chiral environments.

- **Optical Activity**: Enantiomers rotate plane-polarized light in opposite directions. This property is why enantiomers are sometimes called optical isomers. A 50:50 mixture of enantiomers, known as a racemic mixture, does not rotate plane-polarized light because the rotations cancel out.

- **Biological Relevance**: Enantiomers can have drastically different biological effects due to the chiral nature of biological systems. For example, the drug thalidomide has two enantiomers: One is effective for morning sickness, while the other causes severe birth defects. Similarly, the (S)-enantiomer of ibuprofen is more pharmacologically active than the (R)-enantiomer.

- **Chiral Environments**: Enantiomers behave identically in achiral environments but can be distinguished in chiral environments, such as when they interact with other chiral molecules (e.g., enzymes or chiral reagents).

Examples:

- **2-Butanol**: The carbon atom bonded to the -OH group is a chiral center, which results in two enantiomers: (R)-2-butanol and (S)-2-butanol.
- **Amino Acids**: Most naturally occurring amino acids, like alanine, exist as one enantiomer (L-alanine) in biological systems. The mirror image (D-alanine) is rare.

Enantiomers are often distinguished using the R/S system (Cahn-Ingold-Prelog priority rules), which assigns a configuration based on the priority of substituents around the chiral center. Alternatively, the D/L system is used for amino acids and sugars based on their structural similarity to glyceraldehyde.

Diastereomers

Diastereomers are stereoisomers that are not mirror images of each other. They arise in molecules with multiple chiral centers or in systems with restricted rotation (e.g., double bonds or cyclic structures). Unlike enantiomers, diastereomers have different physical and chemical properties, such as melting points, boiling points, solubilities, and reactivities. This makes them easier to separate and identify.

<u>Characteristics of Diastereomers</u>

- **Multiple Chiral Centers**: Diastereomers often occur in molecules with two or more chiral centers, where the stereoisomers differ in the configuration at one or more (but not all) chiral centers. For a molecule with n chiral centers, the maximum number of stereoisomers is 2^n, though the actual number may be lower due to symmetry or meso forms; these stereoisomers include both enantiomers and diastereomers.

- **Different Properties**: Because diastereomers are not mirror images, they interact differently with both achiral and chiral environments. This causes variations in properties like solubility, reactivity, and chromatographic behavior.

This can be exploited for separation techniques like crystallization or chromatography.

- **Geometric Isomers**: A common subtype of diastereomers is the cis-trans isomer (also called a geometric isomer), which occurs in molecules with restricted rotation, such as alkenes (due to their double bonds) or cyclic compounds.

For example, in 2-butene, the cis isomer has methyl groups on the same side of the double bond, while the trans isomer has them on opposite sides. These isomers are diastereomers because they are not mirror images and have different physical properties (e.g., cis-2-butene has a higher boiling point due to a greater dipole moment).

Cis-2-butene   Trans-2-butene

Fig 25: Example of diastereomers.

In cyclic compounds like 1,2-dichlorocyclohexane, the cis isomer (with both chlorine atoms on the same side of the ring) is distinct from the trans isomer (chlorine atoms on opposite sides).

- **Optical Activity**: Diastereomers may or may not be optically active. If a diastereomer has chiral centers, it may rotate plane-polarized light, but the direction and magnitude of rotation differ from other diastereomers. Some compounds, like meso compounds, are achiral despite having chiral centers because they possess an internal plane of symmetry. This renders them optically inactive and not considered diastereomers.

- **Biological Relevance**: Diastereomers can have different biological activities, though their differences are often less dramatic than those of enantiomers. For example, the diastereomers of tartaric acid (a molecule with two chiral centers) have different solubilities and physiological effects.

Examples:

- **Tartaric Acid**: This molecule has two chiral centers, which results in three stereoisomers: (R,R)-tartaric acid, (S,S)-tartaric acid (enantiomers), and the meso compound (achiral due to a plane of symmetry). The (R,R) and (S,S) forms are enantiomers, while each of these and the meso form are diastereomers of one another.

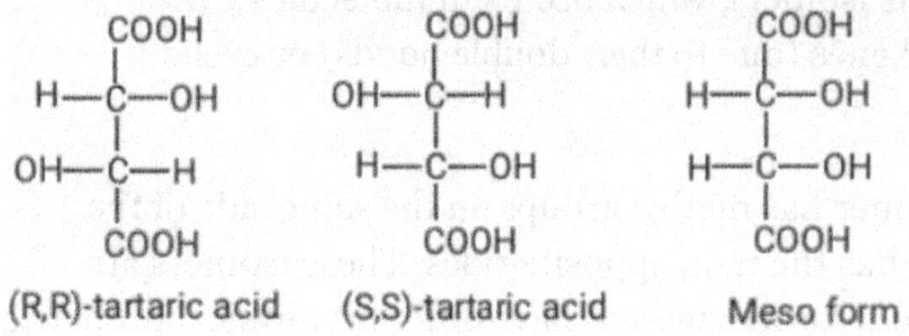

(R,R)-tartaric acid (S,S)-tartaric acid Meso form

Fig 26: Different structures of tartaric acid.

- **1,2-Dibromoethene**: The cis and trans isomers are geometric isomers and a subtype of diastereomers because they are not mirror images and differ in the spatial arrangement of bromine atoms around the double bond.

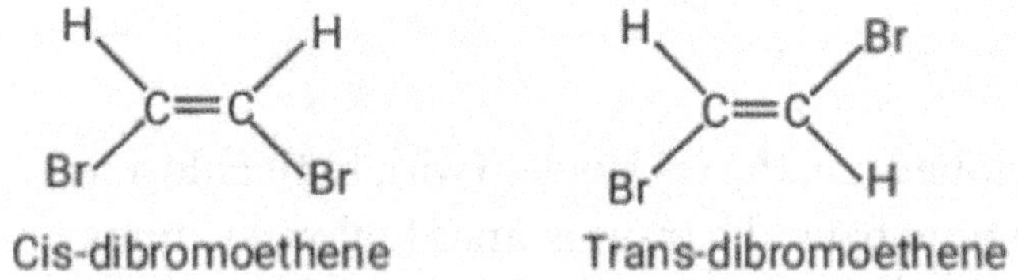

Cis-dibromoethene Trans-dibromoethene

Fig 27: Different structures of dibromoethene.

- **Sugars**: Epimers, a type of diastereomer, differ in configuration at only one chiral center. For example, glucose and galactose are epimers (They differ at the C-4 carbon).

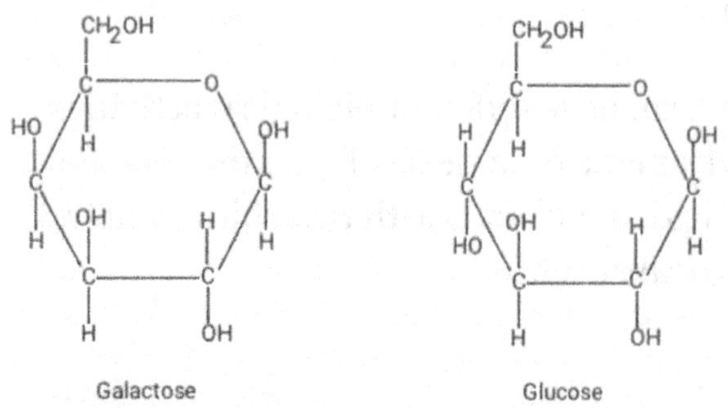

Galactose Glucose

Fig 28: Examples of the structure of epimers (glucose and galactose).

Diastereomers are often named using the R/S system for each chiral center or cis/trans (or Z/E for alkenes, based on Cahn-Ingold-Prelog priority rules). For cyclic compounds or alkenes, cis and trans are commonly used.

Chirality and Optical Activity

Chirality (from the Greek "cheir," which means "hand") refers to molecules that lack a plane or center of symmetry. This makes them nonsuperimposable on their mirror images. Just as left and right hands are distinct, chiral molecules exist as two nonsuperimposable enantiomers.

A plane of symmetry bisects a molecule so that one-half is the mirror image of the other and renders the molecule achiral. A center of symmetry is a point in a molecule such that any line drawn through it encounters identical atoms at equal distances in opposite directions, which renders the molecule achiral.

The most common cause of chirality in organic molecules is a carbon atom bonded to four different substituents, termed a stereogenic center or chiral center.

Optical Activity

Chiral molecules can rotate the plane of plane-polarized light, a property known as optical activity. When plane-polarized light passes through a chiral compound, the plane of polarization is altered.

A chiral compound that rotates light clockwise is termed dextrorotatory and is labeled with a (+) or d prefix. A compound that rotates light counterclockwise is called levorotatory and is designated by a (-) or l prefix.

The extent of rotation depends on the compound's concentration, the light's path length through the sample, the wavelength of light, and temperature. The specific rotation [α] quantifies a compound's ability to rotate plane-polarized light:

- $[\alpha]\lambda T = \alpha / c \cdot l$

Where:

- α is the observed rotation in degrees.
- c is the concentration in g/mL (or g/100 mL, depending on the convention used; be consistent with the system in use).
- l is the path length in decimeters (dm).
- T is the temperature in °C.

- λ is the wavelength of light used (often the sodium D line at 589 nm).

Enantiomers have equal magnitudes of specific rotation but with opposite signs. A racemic mixture contains equal amounts of both enantiomers and results in no net optical rotation, as their effects cancel each other out.

R/S Nomenclature

The Cahn-Ingold-Prelog priority rules are employed to assign configurations (R or S) to stereocenters in chiral molecules. This provides a clear method to describe the three-dimensional arrangement of substituents. The steps include:

- Assign Priorities: Rank the four substituents attached to the chiral center based on the atomic number of the atoms directly bonded to it. Higher atomic numbers receive higher priority (1 > 2 > 3 > 4).
- Handle Isotopes: If two substituents contain the same atom directly bonded to the stereocenter, heavier isotopes (greater mass number) receive higher priority (e.g., deuterium > hydrogen), while branching is resolved by the atomic number of subsequent atoms.
- Treat Multiple Bonds: Multiple bonds are treated as if the atom at the multiple bond is bonded to that many of the atoms at the other end. For instance, a C=O bond is treated as if carbon is bonded to two oxygen atoms.
- Orient the Molecule: Position the molecule so that the lowest priority substituent (4) points away from you.
- Determine the Direction: Examine the order of the remaining three substituents (1, 2, 3). If the sequence from highest to lowest priority is clockwise, the configuration is (R). If counterclockwise, it is (S).

For molecules with multiple chiral centers, each stereocenter is assigned an (R) or (S) configuration independently.

Meso Compounds

Meso compounds are a unique class of achiral molecules, even though they contain chiral centers (stereocenters). This apparent contradiction arises because meso compounds possess an internal plane of symmetry or center of symmetry, which makes them superimposable on their mirror images. As a result, meso compounds are optically inactive, which means they do not rotate plane-polarized light, unlike their chiral counterparts. Meso compounds are significant in stereochemistry because they illustrate how molecular symmetry can override the chirality introduced by stereocenters, which affects the molecule's physical, chemical, and biological properties.

A meso compound is an achiral molecule that contains two or more chiral centers but is rendered achiral due to the presence of a symmetry element, usually a plane of symmetry or, less commonly, a center of symmetry. This symmetry ensures that the molecule is identical to its mirror image and negates the optical activity that would usually arise from chiral centers.

Key Features of Meso Compounds

- Presence of Chiral Centers: Meso compounds have at least two stereocenters, each of which individually contributes to chirality by being bonded to four different substituents (e.g., a tetrahedral carbon atom).
- Internal Symmetry: The molecule possesses a plane of symmetry (or center of symmetry) that bisects it, such that one-half is a mirror image of the other. This symmetry relates the chiral centers in a way that cancels their chiral effects.
- Opposite Configurations: The chiral centers in a meso compound usually have opposite configurations (e.g., one is R, the other is S) and are chemically equivalent (not constitutionally equivalent), meaning they occupy symmetrical positions within the same molecular framework.
- Optical Inactivity: Due to their symmetry, meso compounds do not rotate plane-polarized light, as the rotations from the chiral centers cancel out. This distinguishes them from chiral stereoisomers, which are optically active.
- Achirality: Despite the presence of chiral centers, the molecule as a whole is achiral because it can be superimposed on its mirror image.

Structural Requirements for Meso Compounds

For a molecule to be classified as a meso compound, it must meet specific structural criteria:

- **Two or More Chiral Centers:** Meso compounds require at least two stereocenters, as a single chiral center cannot produce a plane of symmetry that cancels chirality. The number of chiral centers is often even, though this is not strictly necessary if symmetry is present.
- **Plane or Center of Symmetry:** A plane of symmetry bisects the molecule such that one-half is the mirror image of the other. For example, in a molecule with two chiral carbons, the plane may pass through the midpoint between them, relating the R and S centers. A center of symmetry (less common) is a point

where every atom has an identical counterpart equidistant on the opposite side. This is rare in organic meso compounds but can occur in coordination complexes. The symmetry element ensures that the molecule's mirror image is identical to itself, which makes it achiral.

- **Constitutionally Equivalent Stereocenters with Opposite Configurations:** The chiral centers must have identical substituents arranged symmetrically, making them chemically equivalent rather than constitutionally equivalent. The configurations at these centers are opposite (e.g., one R, one S), so their chiral contributions cancel out due to the molecule's symmetry. For example, in a molecule with two chiral carbons, one might be (R) and the other (S), but the plane of symmetry ensures the molecule is achiral.
- **Even Number of Stereocenters (Usual):** Although not an absolute requirement, meso compounds typically have an even number of stereocenters because this arrangement more readily allows for internal symmetry; however, rare exceptions with odd numbers may exist in highly symmetric systems. Odd numbers of chiral centers can result in meso compounds in rare cases, but this is less common.

Examples of Meso Compounds

To illustrate the concept, let's explore some well-known meso compounds and their structural features.

Tartaric Acid (2,3-Dihydroxybutanedioic Acid): Tartaric acid (HOOC-CH(OH)-CH(OH)-COOH) has two chiral centers, each carbon bearing an -OH group, bonded to -H, -OH, -COOH, and the other chiral carbon. Tartaric acid has three stereoisomers due to its two chiral centers ($2^2 = 4$ possible combinations, but symmetry eliminates one pair by making it a meso form, resulting in three distinct stereoisomers):

- (R,R)-Tartaric acid: Chiral, optically active (dextrorotatory).
- (S,S)-Tartaric acid: Chiral, optically active—levorotatory, enantiomer of (R,R).
- Meso-Tartaric acid: Achiral, with one chiral center as (R) and the other as (S).

Meso-tartaric acid has a plane of symmetry that passes through the midpoint of the C-C bond between the chiral carbons and bisects the molecule. This plane relates the (R) and (S) centers, which makes the molecule superimposable on its mirror image.

The meso form is optically inactive because the rotations from the (R) and (S) centers cancel out. Meso-tartaric acid has different physical properties (e.g., melting point, solubility) compared to the chiral (R,R) and (S,S) forms, as it is a diastereomer of the enantiomeric pair.

Examples:

1) 2,3-Dibromobutane (CH_3-CHBr-CHBr-CH_3): has two chiral carbons, each bonded to -H, -Br, -CH_3, and the other chiral carbon. The molecule has three stereoisomers:

- (R,R) and (S,S) enantiomers (chiral, optically active).
- A meso form with (R,S) or (S,R) configurations.

The meso form has a plane of symmetry perpendicular to the C-C bond between the chiral carbons, which makes it achiral. The meso form is optically inactive.

2) 1,2-Dichlorocyclohexane (Cis Isomer): The two chlorine atoms are on the same side of the cyclohexane ring, and the carbons that bear the chlorines are chiral. The molecule has a plane of symmetry that passes through the ring and bisects the C-C bond between the chiral carbons. This makes the cis isomer a meso compound. The trans-1,2-dichlorocyclohexane isomer lacks this symmetry and exists as a pair of enantiomers, which are chiral and optically active.

Distinguish Meso Compounds from Chiral Compounds

Meso compounds must be differentiated from chiral compounds with multiple stereocenters, as the presence of chiral centers alone does not determine chirality. The key distinctions are:

1) Symmetry

- Meso Compounds: Have a plane or center of symmetry, which makes them achiral and superimposable on their mirror images.
- Chiral Compounds: Lack such symmetry, which makes them nonsuperimposable on their mirror images. For example, (R,R)-tartaric acid has no plane of symmetry and is chiral.

2) Optical Activity

- Meso Compounds: Optically inactive due to the cancellation of rotations from opposite-configured chiral centers.
- Chiral Compounds: Optically active, rotating plane-polarized light. For example, (R,R)-tartaric acid is dextrorotatory, and (S,S)-tartaric acid is levorotatory.

3) Number of Stereoisomers

- For a molecule with n chiral centers, the maximum number of stereoisomers is 2^n. Meso compounds reduce this number by introducing symmetry. For example, a molecule with two chiral centers could have up to four stereoisomers (two enantiomeric pairs), but if a meso form exists, there are only three stereoisomers (one meso, one enantiomeric pair). In tartaric acid, the meso form reduces the number of unique stereoisomers from four to three.

4) Physical Properties

- Meso Compounds: As diastereomers of chiral stereoisomers, meso compounds have different physical properties (e.g., melting point, solubility) compared to their chiral counterparts.
- Chiral Compounds: Enantiomers have identical physical properties (except in chiral environments), but diastereomers (including meso compounds) differ.

Examples:

- **Chiral Example:** (R,R)-2,3-butanediol is not chiral, as it is a meso compound due to the presence of a plane of symmetry; thus, it is achiral and optically inactive.
- **Meso Example:** (R,S)-2,3-butanediol is not meso; it does not have a plane of symmetry and is chiral and optically active. The correct meso compound is (R,S)-butane-2,3-diol with symmetric substituents such as (R,S)-tartaric acid.

Conformational Analysis: Newman Projections and Chair Flips

Conformational isomers (or conformers) are different spatial arrangements of atoms resulting from rotation around single bonds. These isomers interconvert rapidly at room temperature and are often not isolable. Conformational analysis studies the relative energies and populations of different conformers.

Newman Projections

A Newman projection visually represents the conformation of a carbon-carbon single bond by looking directly along the bond's axis. The carbon atom closer to the viewer is depicted as a dot with three bonds radiating from it. The further carbon appears as a circle. Its three attached bonds extend from its edge.

For ethane (CH_3-CH_3), two main extreme conformations exist:

- Staggered Conformation: The C-H bonds on the front carbon are maximally spaced from the C-H bonds on the back carbon (dihedral angle of 60°). This conformation has the lowest energy due to minimal torsional strain.
- Eclipsed Conformation: The C-H bonds on the front carbon align directly with those on the back carbon (dihedral angle of 0°), which results in the highest energy conformation due to maximum torsional strain.

Rotation around the C-C bond in ethane creates a potential energy diagram with three energy minima (staggered conformations) and three energy maxima (eclipsed conformations).

For more complex molecules like butane, additional steric interactions between the methyl groups lead to various staggered conformations with differing energies (anti and gauche).

Chair Conformations of Cyclohexane

Cyclohexane (C_6H_{12}) is a common cyclic compound, with its most stable conformation being the chair conformation, which is virtually free from angle and torsional strain. In this conformation, the carbon atoms are arranged in a puckered shape rather than a flat plane.

In the chair conformation, hydrogen atoms (or other substituents) occupy two types of positions:

- Axial (a) Positions: Oriented vertically, pointing up or down relative to the average plane of the ring. There are three up-axial and three down-axial positions.
- Equatorial (e) Positions: Positioned roughly in the plane of the ring, extending outward. The orientation (up or down) of equatorial positions alternates around the ring and corresponds to the orientation of the axial positions on adjacent carbons.

A chair flip (ring flip) describes a conformational change where one chair conformation of cyclohexane transitions to another. During this process, all axial substituents become equatorial, and all equatorial substituents become axial and pass through higher-energy intermediate conformations (e.g., half-chair, twist-boat, boat).

The stability of different chair conformations is affected by substituent size. Larger substituents prefer equatorial positions to minimize 1,3-diaxial interactions, which are steric repulsions between axial substituents on carbons that are one carbon apart. The equatorial position provides more space, which reduces these steric clashes and results in a more stable conformer.

Chapter 5: Nucleophilic Substitution and Elimination Reactions

Nucleophilic substitution and elimination reactions either replace a leaving group with a nucleophile or remove both a leaving group and a hydrogen atom to form a pi bond. Each reaction type can proceed through two main mechanisms: unimolecular (SN1 and E1) and bimolecular (SN2 and E2).

Elimination reactions involve the removal of a small molecule (often water or hydrogen halide) from a saturated compound, which results in the formation of a double bond. The two main mechanisms are E1 and E2.

SN2 Mechanism (Bimolecular Nucleophilic Substitution)

The SN2 reaction occurs in a single step, where the nucleophile attacks the substrate from the back side, opposite to the leaving group, while the leaving group departs. This concerted process causes an inversion of configuration at the carbon undergoing substitution, known as Walden inversion. It occurs in primary and some secondary substrates due to steric hindrance. The reaction rate depends on the concentrations of both the nucleophile and the substrate.

Key Features of SN2:

- Rate Law: The reaction rate is second order, depending on both the substrate and nucleophile concentrations:
- Rate = k[Substrate][Nucleophile]
- Stereochemistry: Inversion of configuration occurs at the attacked stereocenter.
- Substrate Preference: SN2 is favored by primary and unhindered secondary alkyl halides. Tertiary substrates are generally unreactive due to steric hindrance.
- Nucleophile: A strong nucleophile is required, usually an anionic or a neutral nucleophile with a lone pair.
- Leaving Group: A good leaving group is essential (weak base, stable after departure). Common good leaving groups include halides (I^- > Br^- > Cl^-), tosylate (OTs^-), mesylate (OMs^-), and protonated water.
- Solvent Effects: Polar aprotic solvents (e.g., acetone, DMSO, DMF) are preferred as they solvate cations but don't significantly solvate anionic nucleophiles, which makes them more reactive. Polar protic solvents can hinder nucleophilicity by forming hydrogen bonds.

Example: The reaction of hydroxide ion (OH^-) with methyl bromide (CH_3Br) follows the SN2 mechanism, which produces methanol (CH_3OH) and bromide ion (Br^-):

- $OH^- + CH_3Br \rightarrow CH_3OH + Br^-$.

If the carbon with bromine is a stereocenter, the substituted product will have the opposite configuration due to inversion, but methanol (CH_3OH) itself is not chiral and does not have a stereocenter.

SN1 Mechanism (Unimolecular Nucleophilic Substitution)

The SN1 reaction occurs in two steps. The first step is the slow, rate-determining departure of the leaving group, which forms a carbocation intermediate. The second step is the rapid nucleophilic attack on this carbocation.

It usually occurs in tertiary substrates due to carbocation stability. The reaction rate depends only on the concentration of the substrate. The formation of the carbocation can lead to racemization if the nucleophile can attack from either side.

Key Features of SN1:

- Rate Law: The reaction rate is first order, depending only on the substrate concentration:
- Rate = k[Substrate]
- Stereochemistry: This mechanism causes racemization at the stereocenter, as the nucleophile can attack the planar carbocation from either face, which results in a mixture of enantiomers or a slight preference for inversion if the leaving group blocks one side.
- Substrate Preference: SN1 is favored by tertiary and resonance-stabilized secondary alkyl halides that can form stable carbocations. Primary substrates are generally unreactive due to unstable primary carbocations.
- Nucleophile: A weak nucleophile is often sufficient since it does not participate in the rate-determining step, although increased concentration can enhance the reaction rate.
- Leaving Group: A good leaving group is necessary for carbocation formation.
- Solvent Effects: Polar protic solvents (e.g., water, alcohols) stabilize the carbocation intermediate through solvation.

Example: The reaction of 2-bromo-2-methylpropane—($(CH_3)_3CBr$)—with water follows the SN1 mechanism. This results in the formation of 2-methyl-2-propanol—($(CH_3)_3COH$)—and hydrobromic acid (HBr):

- $(CH_3)_3CBr$ (slow) → $(CH3)3C^+ + Br^-$.
- $(CH_3)_3C^+ + H_2O$ (fast) → $(CH_3)_3COH + H_3O^+$.

The intermediate carbocation is planar, which allows water to attack from either side. This causes racemization if the starting material is chiral.

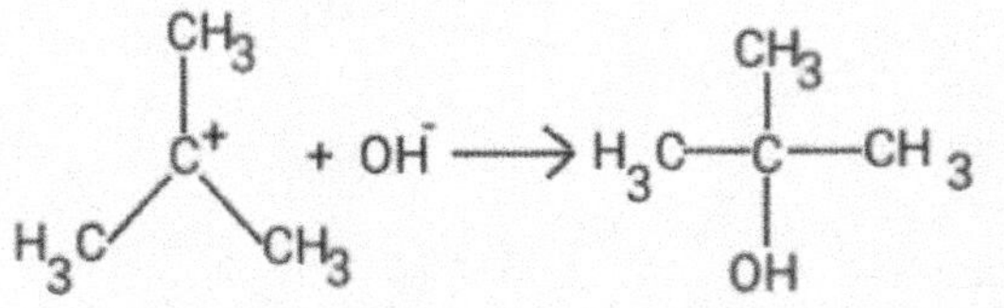

Fig 29: Illustration of SN1 mechanism.

E2 Mechanism (Bimolecular Elimination)

The E2 reaction is a one-step process in which a strong base removes a proton from a carbon adjacent to the one that bears the leaving group. This simultaneously causes the leaving group to depart and form a pi bond (alkene). Effective overlap of orbitals during bond breaking and forming is vital. It is ideally achieved when the hydrogen is removed and the leaving group is in an anti-periplanar arrangement (dihedral angle of 180°).

Key Features of E2:

- Rate Law: The reaction rate is second order, dependent on both substrate and base concentrations.
- Rate = k[Substrate][Base]
- Stereochemistry: The reaction usually favors the formation of the more stable alkene based on Zaitsev's rule, although bulky bases may lead to the formation of less substituted alkenes (Hofmann product). The stereochemistry of the alkene is determined by the substrate's geometry and the required anti-periplanar orientation.
- Substrate Preference: E2 is favored by tertiary and hindered secondary alkyl halides, where stable alkenes can form. Primary substrates can undergo E2, but SN2 is often competitive.
- Base: A strong base (usually anionic, e.g., OH^-, RO^-, $NaNH_2$) is required.
- Leaving Group: A good leaving group is essential.
- Solvent Effects: Polar aprotic solvents favor E2 reactions, as they do not strongly solvate the anionic base.

Example: The reaction of 2-bromopropane with potassium hydroxide (KOH) in ethanol can occur through an E2 mechanism to produce propene ($CH_3CH=CH_2$), potassium bromide (KBr), and water (H_2O):

- $CH_3CHBrCH_3 + OH^- \rightarrow CH_3CH=CH_2 + Br^- + H_2O$.

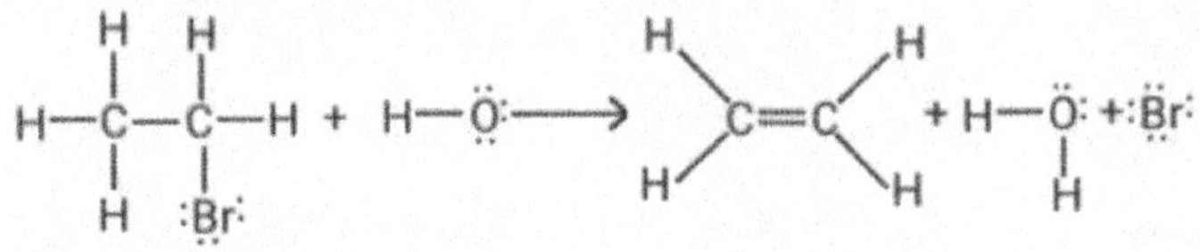

Fig 30: Illustration of the E2 mechanism.

The base removes a β-hydrogen, while the electrons from the C–H bond form a π bond between the α and β carbons, which results in the departure of the bromide ion.

E1 Mechanism (Unimolecular Elimination)

The E1 reaction follows a two-step process similar to SN1. The first step involves the slow, rate-determining departure of the leaving group to form a carbocation intermediate. The second step is the rapid removal of a β-proton by a weak base (often the solvent) to form a pi bond (alkene).

It occurs in tertiary substrates. The reaction rate depends only on the substrate concentration. E1 reactions can lead to both major and minor products due to carbocation rearrangement.

Key Features of E1:

- Rate Law: The reaction rate is first order, depending only on the substrate concentration.
- Rate = k[Substrate]
- Stereochemistry: E1 reactions favor the formation of the more stable alkene (Zaitsev's rule), as the carbocation can lose any adjacent β-proton without a specific coplanar geometry requirement.
- Substrate Preference: E1 is favored by tertiary and resonance-stabilized secondary alkyl halides capable of forming stable carbocations, similar to SN1.
- Base: A weak base (often the solvent, e.g., H_2O, ROH) is used.
- Leaving Group: A good leaving group is necessary for carbocation formation.
- Solvent Effects: Polar protic solvents are preferred as they stabilize the carbocation intermediate.

Example: The reaction of 2-bromo-2-methylpropane in ethanol can undergo E1 elimination to yield 2-methylpropene and hydrobromic acid:

- $(CH_3)_3CBr \rightarrow$ (slow) $(CH_3)_3C^+ + Br^-$.
- $(CH_3)_3C^+ + EtOH \rightarrow$ (fast) $(CH_3)_2C{=}CH_2 + EtOH_2^+$.

The carbocation can lose a proton from one of the methyl groups to form the alkene.

Carbocations

A carbocation is an intermediate species characterized by a positively charged carbon atom that forms three bonds and possesses an empty p orbital. This results in a sextet of electrons instead of the usual octet. This electron deficiency renders carbocations highly reactive intermediates in various organic reactions.

Factors that Influence Carbocation Stability

The stability of carbocations is mainly determined by the degree of alkyl substitution at the positively charged carbon:

- Tertiary (3°) carbocations are the most stable due to significant hyperconjugation and inductive electron donation from three alkyl groups.
- Secondary (2°) carbocations exhibit moderate stability.
- Primary (1°) carbocations are generally unstable and rarely form unless stabilized by resonance or neighboring group participation.
- Methyl carbocations are the least stable.

Additionally, resonance stabilization can enhance carbocation stability if the positive charge can be delocalized within a conjugated system.

Carbocation Rearrangements

Carbocation rearrangements occur to enhance the stability of the intermediate. When a carbocation forms at a less substituted or less stable position, it may rearrange to a more substituted or resonance-stabilized carbocation, which lowers the energy of the intermediate and affects the reaction pathway and products.

The two most common rearrangements include:

- 1,2-Hydride Shifts.
- 1,2-Alkyl (or Alkyl Group) Shifts.

Both types involve the migration of an adjacent atom or group, along with its bonding electrons, to the carbocation center.

1,2-Hydride Shifts

A 1,2-hydride shift entails the migration of a hydrogen atom (along with its bonding electrons) from a carbon adjacent to the carbocation to the positively charged carbon. This migration causes the positive charge to shift to the carbon that lost the hydride.

The migrating hydride is electron-rich and carries both electrons from the C–H bond. The shift is illustrated with a curved arrow from the C–H bond to the carbocation center. The transition state resembles a partial bond formation between the migrating hydrogen and the carbocation carbon, similar to a baton pass in a relay race.

Consider a secondary carbocation next to a secondary or primary carbon with a hydrogen atom. The hydride from the adjacent carbon migrates to the carbocation center, moving the positive charge to the carbon that lost the hydride, provided the rearranged carbocation is more stable (e.g., tertiary). The resulting tertiary carbocation is more stable due to greater alkyl substitution. This rearrangement often occurs spontaneously if a more stable carbocation can be formed, which affects the final product distribution in SN1 and E1 reactions.

1,2-Alkyl Shifts

A 1,2-alkyl shift involves the migration of an alkyl group (such as methyl or ethyl) from a carbon adjacent to the carbocation to the positively charged carbon. The alkyl group moves with its bonding electrons, transferring the carbocation to the carbon it vacated. The curved arrow is drawn from the C–C bond of the migrating alkyl group to the carbocation center. This shift is favored when it causes a more substituted or resonance-stabilized carbocation.

For example, a secondary carbocation adjacent to a tertiary carbon with an alkyl group can undergo an alkyl shift. The alkyl group migrates to the carbocation center, and the positive charge moves to the carbon that lost the alkyl group. The new carbocation is tertiary and more stable. Alkyl shifts are less common than hydride shifts but are important in certain rearrangements, particularly when no suitable hydrogens are available for migration.

Carbocation Rearrangements in SN1 and E1 Reactions

Role in SN1 Reactions

In SN1 reactions, the rate-determining step involves the loss of the leaving group to form a carbocation intermediate. If the initially formed carbocation is not the most stable, it may rearrange before a nucleophilic attack occurs. Rearranged carbocations can lead to products different from those expected by simple substitution at the original carbocation site. For instance, a secondary carbocation formed by the departure of a leaving group can rearrange through a hydride or alkyl shift to form a tertiary carbocation, which then reacts with the nucleophile. This rearrangement explains why SN1 reactions can yield unexpected substitution products.

Role in E1 Reactions

In E1 elimination reactions, the carbocation intermediate can also undergo rearrangement before elimination occurs. The position of the double bond in the alkene product may shift due to carbocation rearrangement. This results in more substituted and stable alkene products. This explains why E1 reactions frequently yield the most substituted alkene, even if it involves a rearrangement.

Rearrangements happen only if a more stable carbocation can be formed. The driving force is the generation of a carbocation with greater alkyl substitution or resonance stabilization. Rearrangements are unlikely in tertiary carbocations, which are already stable, but they are common in secondary carbocations and rarely observed in primary carbocations due to their inherent instability.

How to Identify Potential Rearrangements

Look for a carbocation adjacent to a carbon that has hydrogen atoms or alkyl groups. Assess whether a 1,2-hydride or alkyl shift would yield a more stable carbocation. If so, a rearrangement is likely.

Mechanistic Illustration: SN1 with Hydride Shift

Consider the SN1 reaction of a secondary alkyl halide adjacent to a tertiary carbon:

1) Loss of the leaving group forms a secondary carbocation.

2) A 1,2-hydride shift occurs and moves a hydrogen from the tertiary carbon to the carbocation center.

3) The carbocation is now tertiary and more stable.

4) The nucleophile attacks the carbocation, which produces the substitution product.

5) Deprotonation (if necessary) yields the neutral product.

This sequence explains why the product corresponds to substitution at the tertiary carbon instead of the original secondary carbon.

Mechanistic Illustration: E1 with Alkyl Shift

In an E1 elimination reaction:

1) The leaving group departs and forms a carbocation.

2) A 1,2-alkyl shift occurs if it results in a more stable carbocation.

3) A base removes a β-hydrogen adjacent to the carbocation.

4) A double bond forms, usually at a more substituted position due to rearrangement.

This process clarifies the formation of the more substituted alkene product, consistent with Zaitsev's rule.

Reaction Conditions and Factors

The outcome of nucleophilic substitution or elimination reactions depends on several factors, including the substrate structure, nucleophile/base characteristics, leaving group quality, and solvent type.

Substrate Structure

- Steric Hindrance: Branching near the reaction center can hinder backside attack by the nucleophile in SN2 reactions. The reactivity order for SN2 is usually methyl > primary > secondary >> tertiary. Steric hindrance has less impact on SN1 and E1 reactions, which involve carbocation intermediates.
- Carbocation Stability: Substrates that can form stable carbocations (tertiary or secondary with resonance stabilization) favor SN1 and E1 mechanisms. Primary carbocations are unstable and rarely form unless stabilized by resonance or neighboring group effects, so they are generally not involved in SN1 or E1 reactions.
- Presence of β-Hydrogens: For elimination reactions (E2 and E1) to occur, the substrate must have at least one hydrogen on a carbon adjacent to the leaving

group (β-hydrogen). The number and accessibility of these β-hydrogens influence the rate and regioselectivity of elimination.

Nucleophile/Base

- Strength: Strong nucleophiles promote SN2 reactions, while strong bases favor E2 reactions. The basicity and nucleophilicity of a reagent are related but not identical. For example, bulky strong bases like tert-butoxide are poor nucleophiles and tend to favor elimination (E2) over substitution.
- Concentration: Higher concentrations of a strong nucleophile favor SN2 reactions. The rates of SN1 and E1 are independent of nucleophile/base concentration.
- Size/Bulkiness: Bulky nucleophiles are more likely to cause steric hindrance. They favor elimination (E2) over substitution (SN2), especially with more substituted substrates.

Leaving Group

A good leaving group is important for both substitution and elimination reactions. A good leaving group can depart as a stable, weakly basic species, which carries the electron pair from the cleaved bond. Common good leaving groups include:

- Halides: $I^- > Br^- > Cl^- >> F^-$ (the order may vary in polar protic solvents due to solvation effects).
- Sulfonates: Tosylate (OTs^-), mesylate (OMs^-), and triflate (OTf^-) are excellent leaving groups.
- Water (H_2O) or Protonated Alcohols: These can serve as good leaving groups when the oxygen is protonated.
- Nitrogen Gas (N_2): An exceptionally good leaving group in diazonium salt reactions.

Strongly basic leaving groups (e.g., OH^-, RO^-, NH_2^-) are generally poor leaving groups unless they are protonated (e.g., as H_2O or ROH_2^+) or converted into better leaving groups like sulfonates; otherwise, they are rarely displaced in substitution or elimination reactions under usual conditions.

Solvent

The solvent has a significant impact on the stability of reactants, intermediates, and transition states.

- Polar Aprotic Solvents: Solvents like acetone, DMSO, and DMF have a significant dipole moment but lack acidic hydrogen atoms. They can effectively solvate cations while having weaker interactions with anions. They generally favor SN2 and E2 reactions and keep the anionic nucleophile/base reactive.
- Polar Protic Solvents: Solvents such as water, alcohols, and carboxylic acids can form hydrogen bonds and stabilize both cations and anions. Protic solvents stabilize carbocation intermediates. They favor SN1 and E1 reactions but can also solvate nucleophiles/bases through hydrogen bonding, which may hinder SN2 and E2 reactions.

Competition Between Substitution and Elimination

Both nucleophilic substitution and elimination reactions can occur when an alkyl halide or similar substrate reacts with a nucleophile/base. The reaction pathway depends on various factors, including reaction conditions, substrate characteristics, and reagent nature.

Factors that Favor Substitution

- Strong Nucleophile, Weak Base: Small, highly nucleophilic anions (e.g., I^-, Br^-, HS^-, CN^-) favor substitution, especially in polar aprotic solvents.
- Primary Substrates: Less steric hindrance favors SN2. Primary carbocations are unstable, which makes SN1 unlikely.
- Low Temperatures: Lower temperatures generally favor substitution over elimination due to substitution reactions often having lower enthalpic activation barriers, while elimination is more entropically favored at higher temperatures.

Factors that Favor Elimination

- Strong, Sterically Hindered Base: Bulky bases (e.g., tert-butoxide, DBU, DBN) are poor nucleophiles due to steric hindrance and tend to abstract protons, favoring E2 elimination.
- Tertiary Substrates: Tertiary substrates are more prone to elimination due to steric hindrance against SN2 and the ability to form stable carbocations that can undergo E1.
- High Temperatures: Higher temperatures favor elimination reactions due to a more favorable entropy change ($\Delta S^\circ > 0$ for elimination, as one molecule forms two or more).

Secondary Substrates

Secondary substrates can undergo all four mechanisms (SN1, SN2, E1, E2), and the product distribution is often complex and highly dependent on specific reaction conditions (nucleophile/base strength and size, solvent, temperature).

Table 2: Comparison of SN1, SN2, E1, E2 mechanisms.

Factor	**SN2 Favored By**	**SN1 Favored By**	**E2 Favored By**	**E1 Favored By**
Substrate	Primary < Secondary	Tertiary > Secondary (stabilized C^+)	Tertiary > Secondary (with β-hydrogens)	Tertiary > Secondary (stabilized C^+, β-H)
Nucleophile/Base	Strong nucleophile, weak base, small	Weak nucleophile (solvent)	Strong, often bulky, base	Weak base (solvent)
Leaving Group	Good leaving group	Good leaving group	Good leaving group	Good leaving group
Solvent	Polar aprotic	Polar protic	Polar aprotic	Polar protic
Temperature	Lower	Higher	Higher	Higher

Energy Diagrams and Reaction Coordinates

An energy diagram (or reaction coordinate diagram) illustrates the energy changes that occur during a chemical reaction. It depicts the progression from reactants to products. The x-axis represents the reaction coordinate, which indicates the reaction's progress, while the y-axis shows the potential energy of the system.

Key Features of Energy Diagrams for SN and E Reactions

- Reactants and Products: Represented by energy minima on the diagram. The energy difference between reactants and products (ΔG° for Gibbs free energy or ΔH° for enthalpy) determines whether the reaction is exergonic or endergonic; exergonic reactions have $\Delta G^\circ < 0$, while endergonic reactions have $\Delta G^\circ > 0$.

- Transition State(s): A transition state is a high-energy, unstable species that represents the maximum energy point along the reaction pathway. It corresponds to an energy maximum on the diagram and indicates where bonds form and break.
- Activation Energy (Ea or $\Delta G^\ddagger$): This is the energy difference between the reactants and the transition state. It represents the minimum energy required for the reacting molecules to overcome the energy barrier and proceed to products. Reactions with lower activation energies proceed faster.
- Intermediates (for Multistep Reactions): In reactions like SN1 and E1, intermediates (e.g., carbocations) are formed in one step and consumed in subsequent steps. These intermediates appear as energy minima between two transition states on the energy diagram.

Energy Diagrams for SN2 and E2

SN2 and E2 reactions are single-step, so their energy diagrams show one energy barrier that corresponds to the transition state where bonds break and form simultaneously. The height of this barrier represents the activation energy for the reaction.

Energy Diagrams for SN1 and E1

These reactions occur in two steps. This results in diagrams with two energy barriers and a valley that represents the carbocation intermediate. The first step (carbocation formation) is usually the rate-determining step due to its higher activation energy, while the second step (nucleophilic attack or proton removal) is faster with a lower activation energy.

When we analyze energy diagrams for different mechanisms under various conditions, we can understand why one pathway might be favored over another. Factors that stabilize the transition state of a particular mechanism will lower its activation energy and increase the reaction rate. Similarly, the stability of intermediates in multistep reactions influences the overall reaction rate.

Chapter 6: Electrophilic Addition to Alkenes and Alkynes

Alkenes and alkynes, characterized by their π bonds (areas of high electron density), readily undergo attacks by electrophiles (electron-seeking species). During this process, the π bond breaks, which results in the formation of new σ bonds with the electrophile and another atom or group.

General Mechanism:

1) **Electrophilic Attack:** The electrophile (E^+) approaches the electron-rich π bond and forms a bond with one of the carbon atoms involved in the multiple bonds. This step involves the donation of π electrons to the electrophile, which results in a carbocation intermediate on the other carbon of the multiple bond.

2) **Nucleophilic Attack:** The carbocation is electron-deficient and is subsequently attacked by a nucleophile (Nu^- or a neutral species with a lone pair). This results in the formation of the addition product.

The regioselectivity and stereoselectivity of these reactions are important aspects that will be discussed later.

Hydrohalogenation, Hydration, Hydroboration-Oxidation

These are specific examples of electrophilic addition reactions that occur with alkenes (and sometimes alkynes) and involve different electrophiles and nucleophiles.

Hydrohalogenation

This process involves the addition of hydrogen halides (HX, where X = Cl, Br, I) to alkenes.

- Protonation: The π electrons of the alkene attack the proton (H^+) from the hydrogen halide and form a carbocation on the more substituted carbon (per Markovnikov's rule). The halide ion (X^-) is released as a counterion.
- Nucleophilic Attack: The halide ion (X^-) then attacks the carbocation, which results in the formation of a haloalkane.

Example: The reaction of propene with hydrogen bromide (HBr):

- $CH_3CH=CH_2 + HBr \rightarrow CH_3CH^+CH_3 + Br^- \rightarrow CH_3CHBrCH_3$

This yields 2-bromopropane, which is the Markovnikov product. Carbocation rearrangements (such as 1,2-hydride or alkyl shifts) may occur if a more stable carbocation can be formed.

Hydration

This refers to the addition of water (H_2O) to alkenes to create alcohols, usually facilitated by an acid catalyst.

- **Protonation:** The alkene is protonated by an acid catalyst (e.g., H_3O^+) and forms a carbocation on the more substituted carbon (as per Markovnikov's rule).
- **Nucleophilic Attack:** Water acts as the nucleophile and attacks the carbocation to form a protonated alcohol ($R{-}OH_2^+$).
- **Deprotonation:** A water molecule removes a proton from the protonated alcohol, which yields the neutral alcohol and regenerates the acid catalyst (H_3O^+).

Example: The acid-catalyzed hydration of propene:

- $CH_3CH{=}CH_2 + H_3O^+ \rightleftharpoons CH_3CH^+CH_3 + H_2O$

This causes the formation of 2-propanol, the Markovnikov product. Similar to hydrohalogenation, carbocation rearrangements can occur.

Hydroboration-Oxidation

This two-step process results in the anti-Markovnikov addition of water across an alkene and is stereospecific.

1) **Step 1: Hydroboration**

- Borane (BH_3) is often used as its dimer diborane (B_2H_6) or as a borane-Lewis base complex (e.g., $BH_3{\cdot}THF$). It acts as an electrophile. The boron atom adds to the less substituted carbon of the alkene due to steric factors and the partial positive charge on the more substituted carbon during the transition state. The hydrogen atom from BH_3 adds to the more substituted carbon, which results in syn addition (both boron and hydrogen add to the same face of the alkene). This process repeats until all three B-H bonds of borane react with alkene molecules.

2) **Step 2: Oxidation**

- The trialkylborane intermediate is then treated with hydrogen peroxide (H_2O_2) in a basic solution (e.g., NaOH). The boron atom is replaced by a hydroxyl group (OH) and maintains the stereochemistry established during hydroboration.

Overall Reaction:

- $3RCH{=}CH_2 + BH_3 \rightarrow (RCH_2CH_2)_3B \rightarrow [H_2O_2, NaOH]\ 3RCH_2CH_2OH$

Example: Hydroboration-oxidation of propene:

$$CH_3CH = CH_2 \xrightarrow{1.BH_3\cdot THF \quad 2.H_2O_2,NaOH} CH_3CH_2CH_2OH$$

This results in 1-propanol, the anti-Markovnikov product, with syn addition of H and OH groups.

Regioselectivity and Stereoselectivity

These terms describe the preference for forming one constitutional isomer or stereoisomer over others in a chemical reaction.

Regioselectivity

Regioselectivity refers to the preference for bond formation at a specific atom within a molecule. In addition, in reactions that involve alkenes and alkynes, regioselectivity often determines which carbon atom of the multiple bond the electrophile and nucleophile will attach to.

- **Markovnikov's Rule:** In the addition of HX or H_2O (acid-catalyzed) to an alkene, the hydrogen atom from the reagent adds to the carbon with more hydrogen atoms, while the halogen or hydroxyl group adds to the more substituted carbon. This is because the reaction proceeds through the more stable carbocation, usually the one with more alkyl groups attached.

- **Anti-Markovnikov Addition:** Reactions that proceed with opposite regioselectivity to Markovnikov's rule are termed anti-Markovnikov additions. Hydroboration-oxidation exemplifies this, with hydrogen from BH_3 adding to the more substituted carbon and the hydroxyl group from oxidation ending up on the less substituted carbon. Radical addition of HBr to alkenes in the presence of peroxides also exhibits anti-Markovnikov regioselectivity.

Stereoselectivity

Stereoselectivity refers to the preference for forming one stereoisomer over other possible stereoisomers when multiple stereoisomers can be produced.

- **Syn Addition:** This occurs when two atoms or groups add to the same face of the alkene or alkyne. Hydroboration is a syn addition, as is catalytic hydrogenation (addition of H_2 with a metal catalyst).

- **Anti Addition:** This happens when two atoms or groups add to opposite faces of the alkene or alkyne. Bromination of alkenes follows an anti-addition mechanism that involves a cyclic bromonium ion intermediate.

- **Stereospecificity:** A reaction is stereospecific if the stereochemistry of the reactant dictates the stereochemistry of the product. For example, the SN2 reaction is stereospecific, as a chiral substrate will always undergo inversion of configuration. Hydroboration-oxidation is also stereospecific, as the syn addition of boron and hydrogen to the alkene determines the stereochemistry of the intermediate, and the oxidation step retains this stereochemistry in the resulting alcohol.

Addition to Carbonyls: Aldehydes and Ketones

The carbonyl group (C=O) in aldehydes and ketones is polar, with a partial positive charge on the carbon atom and a partial negative charge on the oxygen atom. This polarity renders the carbonyl carbon electrophilic and susceptible to nucleophilic attack. The oxygen atom, with its lone pairs, can also act as a Lewis base and be protonated by acids.

General Mechanism of Nucleophilic Addition to Carbonyls:

1) **Nucleophilic Attack:** A nucleophile attacks the electrophilic carbonyl carbon. The π bond between carbon and oxygen breaks, which transfers electrons to the oxygen atom and creates an alkoxide intermediate with a negative charge on the oxygen.

2) **Protonation:** Under acidic or neutral conditions, the alkoxide intermediate is usually protonated by the solvent or acid and yields a neutral alcohol derivative.

Common nucleophiles that add to carbonyls include:

- **Hydride reagents:** Sodium borohydride ($NaBH_4$) and lithium aluminum hydride ($LiAlH_4$) reduce aldehydes to primary alcohols and ketones to secondary alcohols. The hydride ion (H^-) serves as the nucleophile.
- **Grignard reagents (RMgX) and organolithium reagents (RLi):** These strong carbon nucleophiles react with carbonyls to form new carbon-carbon bonds. Aldehydes yield primary alcohols, while ketones yield secondary alcohols upon protonation of the alkoxide intermediate.
- **Alcohols (in the presence of acid catalyst):** Alcohols can be added to aldehydes to form hemiacetals and acetals (if two equivalents of alcohol are used). Ketones form hemiketals and ketals, which serve as important protecting groups for aldehydes and ketones.
- **Amines:** Primary amines (RNH_2) react with aldehydes and ketones to produce imines (Schiff bases) with the elimination of water, while secondary amines (R_2NH) yield enamines.
- **Hydrogen cyanide (HCN):** This adds to aldehydes and ketones to form cyanohydrins, which are useful intermediates in organic synthesis. The cyanide ion (CN^-) acts as the nucleophile.

Stereochemistry of Carbonyl Addition

When the carbonyl carbon becomes a stereocenter after nucleophilic addition, the product's stereochemistry depends on the chirality of the original aldehyde or ketone and the nature of the nucleophile and reaction conditions. If the starting carbonyl is not chiral, the addition of an achiral nucleophile can lead to a racemic mixture if a new stereocenter is formed.

Addition to α,β-Unsaturated Carbonyl Compounds

Carbonyl compounds with a double bond adjacent to the carbonyl group (α,β-unsaturated carbonyls) can undergo two types of nucleophilic addition.

1) **1,2-Addition (Direct Addition):** The nucleophile directly attacks the carbonyl carbon.
2) **1,4-Addition (Conjugate Addition or Michael Addition):** The nucleophile attacks the β-carbon of the α,β-unsaturated system. This occurs because the π electrons of the double bond and the carbonyl group are conjugated, and the β-carbon exhibits a partial positive charge due to resonance.

The regioselectivity of nucleophilic addition to α,β-unsaturated carbonyls depends on the nature of the nucleophile and the reaction conditions. Strong, hard nucleophiles (e.g., Grignard reagents, organolithium reagents, and hydride reagents) tend to favor 1,2-addition, while softer, more stabilized nucleophiles (e.g., enolates, cyanide, thiols, amines) usually favor 1,4-addition.

Chapter 7: Spectroscopy in Organic Chemistry

Spectroscopy is a vital technique in organic chemistry that is used to determine molecular structures. It enables us to "visualize" molecules at an atomic level through an examination of their interactions with various forms of electromagnetic radiation and an analysis of how they fragment.

Infrared (IR) Spectroscopy

IR spectroscopy relies on the principle that molecules absorb specific frequencies of infrared radiation that correspond to the vibrational frequencies of their bonds. When the frequency of IR radiation matches a bond's vibrational frequency, the bond absorbs energy, which increases its vibrational amplitude. If we analyze the absorbed frequencies, this allows us to identify the functional groups present in a molecule.

- Wavenumber ($\bar{\nu}$): IR spectra are usually displayed as transmittance or absorbance against wavenumber, defined as the number of waves per centimeter (cm^{-1}). Wavenumber is directly proportional to frequency ($\bar{\nu} = \nu/c$, where $\bar{\nu}$ is the wavenumber and ν is frequency), and energy is given by $E = h\nu = hc\bar{\nu}$, where h is Planck's constant and c is the speed of light.
- Molecular Vibrations: Bonds in molecules vibrate in various modes, such as stretching (change in bond length) and bending (change in bond angle). Different bond types (e.g., C-H, C=O, O-H) and vibrational modes absorb IR radiation at specific frequencies.
- Fingerprint Region (below 1500 cm^{-1}): This complex region contains many overlapping peaks due to various bending and skeletal vibrations. It is unique to each molecule and can be used to identify specific compounds by comparing spectra with a database.
- Functional Group Region (above 1500 cm^{-1}): This area generally has fewer, more distinct peaks that correspond to stretching vibrations of common functional groups. This makes it efficient for identifying specific functional groups.

Table 3: Characteristic IR Absorptions

Functional Group	**Bond**	**Wavenumber Range (cm^{-1})**	**Intensity**	**Appearance**
Alkanes	C-H stretch	2,850–3,000	Medium	Multiple peaks
Alkenes	C=C stretch	1,620–1,680	Weak to Medium	Sharp
Alkynes	C≡C stretch	2,100–2,260	Weak to Medium	Sharp
Aromatic Rings	C=C stretch	1,500–1,600, ~1,600, ~1,450	Weak to Medium	Multiple peaks
Alcohols, Phenols	O-H stretch	3,200–3650	Strong	Broad
Carboxylic Acids	O-H stretch	2,500–3,300	Strong	Very broad
Aldehydes	C=O stretch	1,720–1,740	Strong	Sharp
Ketones	C=O stretch	1,705–1,725	Strong	Sharp

Factors that Affect IR Frequencies

- Bond Strength: Stronger bonds vibrate at higher frequencies (e.g., C≡C > C=C > C–C).
- Atomic Mass: Lighter atoms vibrate at higher frequencies (e.g., C–H > C–D, O–H > O–D).
- Hybridization: The hybridization state of carbon influences the C–H stretching frequency ($sp^3 < sp^2 < sp$).
- Inductive and Resonance Effects: EWGs can increase bond strength and vibrational frequency. Resonance effects can either raise or lower frequencies depending on bond order.
- Hydrogen Bonding: Intermolecular hydrogen bonding can broaden and shift O-H and N-H stretching frequencies to lower wavenumbers.

NMR Spectroscopy: Proton and Carbon NMR

NMR spectroscopy utilizes the magnetic properties of atomic nuclei. Nuclei with an odd number of protons or neutrons (e.g., ^1H, ^{13}C) exhibit nuclear spin and generate a magnetic moment. When placed in an external magnetic field, these nuclei can align with or against the field. Absorption of radio-frequency radiation can induce a "flip" in the spin state, and the frequencies at which this occurs provide insight into the magnetic environment of the nuclei within the molecule.

Proton NMR (^1H NMR)

^1H NMR spectroscopy reveals information about the number and types of hydrogen atoms in a molecule and their connectivity.

Key Concepts:

- Chemical Shift (δ): The resonance frequency of a proton is affected by the surrounding electron density. EWGs deshield protons and cause them to resonate at higher frequencies (higher ppm values), while EDGs shield protons, which results in lower frequencies. Chemical shifts are reported in parts per million (ppm) relative to tetramethylsilane (TMS), defined as 0 ppm.
- Integration: The area under each peak in a ^1H NMR spectrum correlates with the number of equivalent protons that contribute to that signal and allows for the determination of the relative amounts of different types of hydrogen atoms.
- Spin-Spin Splitting (Coupling): The magnetic field experienced by a proton can be influenced by the spin of nearby nonequivalent protons, which results in signal

splitting into multiple peaks (multiplets). The number of peaks follows the n+1 rule, where *n* represents the number of equivalent neighboring protons.

- Coupling Constant (J): The distance between peaks in a multiplet, measured in hertz (Hz), is the coupling constant. Its magnitude depends on the number of bonds (typically 2–3) between coupled protons and, in some cases (e.g., vicinal coupling), on the dihedral angle between them.

Table 4: The usual ^1H NMR Chemical Shifts (ppm)

Type of Proton	**Chemical Shift (δ)**
Alkane ($R-CH_3$, $R-CH_2-R$)	0.8–1.7
Allylic ($R_2C=C-CH_3$)	1.6–1.9
Benzylic ($Ar-CH_3$)	2.2–3.0
α to carbonyl ($RCOCH_3$)	2.0–2.7
α to halogen (RCH_2X)	3.1–4.1
α to oxygen (RCH_2OR)	3.3–4.0
Vinylic ($R_2C=CH_2$)	4.5–6.0
Aromatic ($Ar-H$)	6.5–8.0
Aldehyde ($RCHO$)	9.5–10.5
Carboxylic acid ($RCOOH$)	10–13
Alcohol (ROH)	1–5 (variable)
Amine (RNH_2, R_2NH)	1–5 (variable)

Carbon NMR (^{13}C NMR)

^{13}C is a less abundant isotope of carbon (about 1.1%) and has lower sensitivity in NMR compared to ^1H. However, ^{13}C NMR provides direct information about a molecule's carbon skeleton.

- Chemical Shift (δ): Similar to ^{1}H NMR, the chemical shift of a ^{13}C nucleus is influenced by its electronic environment, but the range is larger (usually 0-220 ppm).
- Number of Signals: Each unique carbon atom in a molecule produces a distinct signal in the ^{13}C NMR spectrum. Equivalent carbons (due to symmetry) generate a single signal.
- Splitting: In simple ^{13}C NMR spectra (proton-decoupled), signals appear as singlets because the coupling between ^{13}C nuclei and protons is removed through broadband decoupling. This simplifies the spectrum and makes it easier to count nonequivalent carbon atoms. Techniques like DEPT (Distortionless Enhancement by Polarization Transfer) can be employed to determine the number of hydrogens attached to each carbon.

Table 5: The usual ^{13}C NMR Chemical Shifts (ppm)

Type of Carbon	**Chemical Shift (δ)**
Alkane	0–30
Alkyl halide	15–40
Ether, Alcohol	50–90
Alkene	100–150
Aromatic	110–160
Alkyne	65–90
Carbonyl (Ketone, Aldehyde)	190–220
Carbonyl (Ester, Acid, Amide)	165–185

Nitrile	115–130

Mass Spectrometry: Fragmentation Patterns

Mass spectrometry (MS) is an analytical technique used to measure the mass-to-charge ratio (m/z) of ions. A molecule is ionized, and these ions are separated based on their m/z values and detected. Mass spectrometry provides insights into the molecular weight of a compound and its fragmentation patterns, which can reveal structural information.

- Molecular Ion (M^+): The ion formed by the removal of one electron from the parent molecule. The m/z value of the molecular ion corresponds to the compound's molecular weight. The molecular ion peak may not always be prominent or even observed, especially for unstable molecules.
- Isotopic Peaks (M+1, M+2, ...): The presence of isotopes (e.g., ^{13}C, ^{2}H, ^{15}N, ^{17}O, ^{18}O, ^{37}Cl, ^{81}Br) causes peaks at m/z values higher than the molecular ion peak. The relative intensities of these isotopic peaks can provide information about the elemental composition of the molecule, such as the presence and number of chlorine or bromine atoms.
- Fragmentation: The molecular ion is often unstable and can break apart into smaller ions and neutral fragments. The resulting pattern of fragment ions is characteristic of the molecule's structure and can be used to deduce structural features.
- Base Peak: The most abundant ion in the mass spectrum (the tallest peak) is termed the base peak, with its relative abundance defined as 100%.

Common Fragmentation Pathways

Fragmentation usually occurs at the weakest bonds or through pathways that result in relatively stable ions (e.g., resonance-stabilized carbocations). Common fragmentation patterns include:

- α-Cleavage: Occurs next to heteroatoms or π systems and forms resonance-stabilized ions.
- Homolytic Cleavage: Symmetrical bond breaking forms two radicals, one of which becomes an ion.
- Heterolytic Cleavage: Unsymmetrical bond breaking forms a cation and an anion (usually, only the cation is observed).

- Rearrangements: Some fragmentations involve atom rearrangements, such as the McLafferty rearrangement, which is common in molecules with a γ-hydrogen and a carbonyl group.

Example:

For a ketone like 2-pentanone ($CH_3COCH_2CH_2CH_3$), common fragmentation pathways include cleavage of bonds adjacent to the carbonyl group. This results in ions like CH_3CO^+ (m/z 43) and $CH_3CH_2CH_2^+$ (m/z 43 is incorrect for this ion; it should be m/z 43 for CH_3CO^+, and $CH_3CH_2CH_2^+$ corresponds to m/z 43, but they are different ions). $CH_3COCH_2^+$ (m/z 71) and CH_3^+ (m/z 15) are also observed, depending on fragmentation. The relative abundances of these fragment ions depend on their stability.

General Approach to Spectral Interpretation

1) Molecular Formula (from Mass Spectrometry and/or Elemental Analysis): Determine the molecular formula, if possible. Calculate the degree of unsaturation (number of rings and/or π bonds).

2) Mass Spectrometry: Identify the molecular ion peak to confirm molecular weight. Look for isotopic peaks (e.g., for Cl or Br). Analyze the fragmentation pattern for insights into structural units.

3) IR Spectroscopy: Identify major functional groups based on characteristic absorptions in the functional group region (above 1500 cm^{-1}). Note the presence or absence of key peaks (e.g., O-H, C=O, N-H, C≡N, C≡C, C=C).

4) NMR Spectroscopy (^1H and ^{13}C):

- ^{13}C NMR: Determine the number of nonequivalent carbon atoms and the types of carbons present (based on chemical shifts and DEPT data, if available).
- ^1H NMR: Assess the number of different types of protons based on signals and their integration. Use chemical shifts to identify the electronic environment of each proton type. Analyze spin-spin splitting patterns to deduce proton connectivity and identify neighboring protons.

5) Propose a Structure: Based on the information from all spectra, suggest one or more possible structures consistent with the data.

6) Verify the Structure: Check if the proposed structure correlates with the observed spectral features. Predict the expected IR absorptions, NMR chemical shifts, splitting patterns, and major mass spectral fragments for the proposed structure, comparing them with the given data.

7) Refine or Revise: If the proposed structure does not match the spectral data, reconsider the interpretation of the spectra and suggest alternative structures.

Chapter 8: Radical Reactions in Organic Chemistry

Radical reactions introduce a new category of reactive intermediates known as radicals, which are species that contain an unpaired electron. These reactions usually proceed through chain mechanisms and exhibit distinct characteristics. This section will explore radical chain mechanisms, the halogenation of alkanes, radical additions to alkenes, and their significance in both industrial and biological contexts.

Radical Chain Mechanisms: Initiation, Propagation, Termination

Radical reactions often follow a chain mechanism, which consists of three main stages: initiation, propagation, and termination.

1) Initiation: This step involves the generation of radicals, usually from a stable precursor molecule. Initiation can be triggered by:

- Homolytic Cleavage: A covalent bond is broken in such a way that each atom retains one of the shared electrons. This can occur through heat (thermolysis), light (photolysis), or the introduction of a radical initiator. Common initiators include peroxides (e.g., benzoyl peroxide, which decomposes to form radicals), azo compounds (e.g., AIBN), and halogens under UV light.

Example: The thermal decomposition of benzoyl peroxide generates benzoyloxy radicals.

2) Propagation: In this stage, radicals react with stable molecules to generate new radicals, which continues the chain reaction. The overall number of radicals remains approximately constant during these steps, as each propagation step typically generates a new radical to replace the one consumed.

Example: In the radical chlorination of methane (CH_4):

- A chlorine radical abstracts a hydrogen atom from methane:

$$Cl^{\bullet} + CH_4 \rightarrow HCl + CH_3^{\bullet}$$

- The methyl radical then reacts with a chlorine molecule to form chloromethane and another chlorine radical:

$$CH_3^{\bullet} + Cl_2 \rightarrow CH_3Cl + Cl^{\bullet}$$

The chlorine radical can react with another methane molecule and perpetuate the chain reaction.

3) Termination: This stage involves the removal of radicals, usually through the combination of two radicals or their reaction with an inhibitor. Termination reduces the concentration of radicals and eventually halts the chain reaction.

Example: Possible termination steps in the radical chlorination of methane include:

- $Cl^{\bullet} + Cl^{\bullet} \rightarrow Cl_2$
- $CH_3^{\bullet} + CH_3^{\bullet} \rightarrow C_2H_6$
- $CH_3^{\bullet} + Cl^{\bullet} \rightarrow CH_3Cl$

The overall rate of a radical chain reaction is influenced by the relative rates of the initiation, propagation, and termination steps. Efficient chain reactions have long kinetic chain lengths. This allows a single initiation event to lead to multiple propagation cycles before termination occurs.

Halogenation of Alkanes

The reaction of alkanes with halogens (Cl_2 or Br_2) in the presence of light or heat proceeds through a radical chain mechanism, which results in the substitution of one or more hydrogen atoms with halogen atoms.

Mechanism (Using Chlorination as an Example):

- Initiation: Homolytic cleavage of Cl_2 by UV light:

$$Cl_2 \xrightarrow{h\nu} 2Cl^{\bullet}$$

- Propagation:
  1) A chlorine radical abstracts a hydrogen atom from the alkane:

$$Cl^{\bullet} + RH \rightarrow HCl + R^{\bullet}$$

  2) The alkyl radical reacts with a chlorine molecule:

$$R^{\bullet} + Cl_2 \rightarrow RCl + Cl^{\bullet}$$

- Termination: Radicals combine:
  - $Cl^{\bullet} + Cl^{\bullet} \rightarrow Cl_2$
  - $R^{\bullet} + R^{\bullet} \rightarrow R\text{-}R$
  - $R^{\bullet} + Cl^{\bullet} \rightarrow RCl$

Regioselectivity

Halogenation of alkanes is generally not highly regioselective and often yields a mixture of products when alkanes with different hydrogen types (primary, secondary, tertiary) react. However, the reactivity order of hydrogen atoms is tertiary > secondary > primary, reflecting the stability of the resulting alkyl radical. Bromination is more selective than chlorination due to the higher activation energy and greater transition state sensitivity in bromination, which leads to stronger preference for forming the more stable (tertiary) radical.

Additional Factors that Influence Regioselectivity

Several factors beyond radical stability influence regioselectivity. Temperature affects selectivity, as higher temperatures provide energy to overcome transition state differences. They make all positions more reactive and reduce selectivity, while lower temperatures enhance selectivity, particularly for bromination. The type of halogen is also key: Fluorination is extremely exothermic and highly reactive, often leading to uncontrollable reactions and complex mixtures, making it difficult to use selectively in alkane halogenation, while iodination is endothermic and impractical due to the iodine radical's low reactivity. In solution-phase reactions, solvents may stabilize radicals or transition states, though gas-phase halogenation, common for alkanes, is less affected. Substituents can modulate reactivity: EWGs like $-CF_3$ destabilize radicals at nearby carbons and reduce reactivity, while allylic or benzylic hydrogens are highly reactive due to resonance stabilization, which enhances selectivity in molecules like toluene, while bromination favors the benzylic position ($C_6H_5CH_2Br$).

Stereochemistry

When an alkane substrate possesses chirality, radical halogenation at a stereocenter can result in racemization. This occurs because the intermediate alkyl radical formed during the reaction is planar or undergoes rapid inversion, which enables the halogen atom to attack from either side of the molecule. As a consequence, two enantiomers are produced in equal amounts, which results in a racemic mixture. This phenomenon highlights the importance of stereochemical considerations in radical reactions, particularly when the substrate includes a stereocenter. The planar nature of the radical intermediates means that the spatial arrangement of substituents can change, which affects the optical activity of the final products.

Despite its utility, radical halogenation presents several limitations that can complicate the synthesis and purification of products.

One significant challenge is polysubstitution, where multiple hydrogen atoms in the alkane can be replaced by halogens. This results in a complex mixture of products, each with varied degrees of substitution. The presence of multiple reactive sites increases the likelihood of forming a complex mixture of mono-, di-, and polyhalogenated products, which makes it difficult to isolate the desired compound. The complexity of these mixtures can complicate subsequent purification processes, as traditional separation methods may not effectively discriminate between closely related constitutional isomers.

Another limitation is the lack of high regioselectivity, particularly in the case of chlorination. Chlorination often causes the formation of a variety of constitutional isomers due to the comparable reactivity of chlorine with primary, secondary, and tertiary C–H bonds, especially under elevated temperatures. This nonselective behavior can result in mixtures that contain a wide array of different isomers, which complicates the isolation of any specific product. The uneven distribution of products can be problematic in synthetic applications where a particular isomer is desired for its unique properties or biological activity.

Radical Additions to Alkenes

Radicals can also add to the π bond of alkenes, which generates a new radical that can propagate a chain reaction.

General Mechanism:

1) Initiation: Generation of a radical initiator (X^{-}).

2) Propagation:

- The radical adds to the alkene and forms a new carbon-centered radical:

$$X^{\bullet} + CH_2 = CHY \rightarrow X\text{-}CH_2 - \dot{C}Y$$

The new radical reacts with another molecule and transfers the radical character. Finally, it forms a new product:

$$\text{X-CH}_2 - \dot{C}Y + \text{Z-Y} \rightarrow \text{X-CH}_2\text{CH}YZ^{\bullet}$$

3) Termination: Combination of radicals.

Hydrohalogenation of Alkenes (Anti-Markovnikov)

In the presence of peroxides, the addition of HBr to alkenes occurs through a radical mechanism and displays anti-Markovnikov regioselectivity, with the bromine atom adding to the less substituted carbon. HCl and HI usually do not undergo anti-Markovnikov addition under these conditions because the corresponding radical chain propagation steps are either not energetically favorable (HCl) or do not proceed efficiently (HI) due to the weakness of the I–H bond and instability of iodine radicals.

Mechanism of Anti-Markovnikov HBr Addition:

1) Initiation: Peroxides decompose to form alkoxy radicals, which abstract a hydrogen atom from HBr and generate bromine radicals:

$$\text{ROOR} \rightarrow 2\text{RO}^{\bullet}$$

$$\text{RO}^{\bullet} + \text{HBr} \rightarrow \text{ROH} + \text{Br}^{\bullet}$$

2) Propagation:

- The bromine radical adds to the alkene (bromine attaches to the less substituted carbon and forms a more stable alkyl radical):

$$\text{Br}^{\bullet} + \text{CH}_3\text{CH} = \text{CH}_2 \rightarrow \text{CH}_3 - \dot{C}H\text{CH}_2\text{Br}$$

- The alkyl radical abstracts a hydrogen atom from another HBr molecule:

$$\text{CH}_3 - \dot{C}H\text{CH}_2\text{Br} + \text{HBr} \rightarrow \text{CH}_3\text{CH}_2\text{CH}_2\text{Br} + \text{Br}^{\bullet}$$

3) Termination: Combination of radicals.

Polymerization of Alkenes

Radical addition to alkenes is the basis for the polymerization of important monomers like ethylene (to polyethylene), styrene (to polystyrene), vinyl chloride (to polyvinyl chloride), and methyl methacrylate (to Plexiglas). A radical initiator adds to a monomer unit to start the chain reaction. This generates a new radical that can then react with additional monomers. This continues until termination occurs.

Other Radical Additions

Radicals can also add to alkynes, and some species (e.g., thiols, RSH) can undergo radical addition to alkenes, often following anti-Markovnikov or non-regioselective pathways, depending on conditions and the specific radical mechanism involved.

Industrial and Biological Examples

Radical reactions play vital roles in various industrial processes and biological systems.

Industrial Examples

Polymer Production

Radical polymerization is a cornerstone of the polymer industry. It produces a wide array of materials like polyethylene, polystyrene, and polyvinyl chloride (PVC). This process initiates a chain reaction with radicals, often generated by peroxides or azo compounds. This adds to monomer units and forms long polymer chains. The versatility of radical polymerization allows the creation of polymers with tailored properties, such as flexibility for plastic bags or rigidity for pipes. This makes it essential for the packaging, construction, and automotive sectors.

The advantages include its robustness and compatibility with a variety of monomers. Innovations like controlled radical polymerization (e.g., atom transfer radical polymerization, ATRP) enable precise control over polymer architecture and produce advanced materials like self-healing coatings. However, challenges include managing heat generation during exothermic reactions and minimizing side reactions. Green chemistry is driving the development of water-based or solvent-free radical polymerization to reduce environmental impact, which ensures sustainable production of next-generation polymers.

Halogenation of Alkanes

Radical halogenation of alkanes is usually initiated by light or heat. It introduces halogen atoms (e.g., chlorine or bromine) into hydrocarbons and produces halogenated compounds like chloroform or chloromethane. These serve as intermediates for solvents, refrigerants, and agrochemicals. For example, chlorofluorocarbons (CFCs) were once widely used as refrigerants and were typically synthesized via multistep processes involving halogen exchange or addition reactions, with radical chlorination sometimes used in precursor steps; their environmental impact eventually led to their phase-out.

The process is valued for its simplicity and ability to functionalize inert alkanes. However, it often lacks selectivity and produces mixtures of products that require purification. Advances in photocatalysis, using light-driven radical initiators, have improved selectivity and energy efficiency. The shift toward greener alternatives, such as bio-based halogenated compounds, addresses environmental concerns, particularly for applications in pharmaceuticals and specialty chemicals.

Cracking of Petroleum

Petroleum cracking employs high-temperature radical reactions to break large hydrocarbon molecules into smaller, more valuable fractions, such as gasoline and kerosene. In thermal cracking, radicals form when carbon-carbon bonds cleave at high temperatures and initiate chain reactions that yield lighter hydrocarbons. This process is important to meet global fuel demands and produce petrochemical feedstocks.

The benefits include the ability to convert heavy crude oil into usable products. However, cracking is energy-intensive and generates byproducts like coke, which can foul reactors. Catalytic cracking primarily involves carbocation (ionic) mechanisms, not radical mechanisms, and improves efficiency and selectivity compared to thermal cracking. New trends involve the integration of renewable feedstocks, like biomass, into cracking processes to produce bio-based fuels. This aligns well with sustainability goals. Innovations in reactor design and catalysts reduce energy use, which makes cracking more environmentally friendly.

Antioxidants

Antioxidants include compounds such as hindered phenols and aromatic amines. They are radical scavengers that inhibit unwanted radical chain reactions, like oxidation, in polymers, fuels, and foods. In polymers, antioxidants prevent degradation from UV exposure or heat and extend the life span of products like plastic packaging. In fuels, they inhibit gum formation and ensure stability during storage. In food, compounds like butylated hydroxyanisole protect against rancidity.

The use of antioxidants enhances product durability and safety. Synthetic antioxidants are highly effective, but concerns about their environmental persistence have spurred research into bio-based alternatives, such as tocopherols (vitamin E) or plant-derived polyphenols. Challenges include how to balance antioxidant efficacy with cost and also ensure compatibility with diverse materials. Advances in nanotechnology, like antioxidant-loaded nanoparticles, have improved delivery and performance, particularly in high-performance polymers and coatings.

Biological Examples

Free Radical Damage

Free radicals, such as superoxide anion ($O_2^{\bullet -}$), hydroxyl radical (•OH), and lipid peroxyl radicals (ROO•), are generated during metabolic processes or through exposure to radiation, pollutants, or toxins. These highly reactive species can damage DNA, proteins, and lipids and contribute to aging and diseases like cancer, cardiovascular disorders, and neurodegenerative conditions. For example, lipid peroxidation in cell membranes, driven by ROO•, disrupts cellular integrity and accelerates disease progression.

Radicals are a natural byproduct of metabolism, particularly in mitochondria during respiration. Better knowledge of their role has spurred research into how to mitigate their effects. Challenges include how to detect and quantify radical damage in vivo. Advances in imaging techniques, like electron paramagnetic resonance, and biomarkers of oxidative stress have improved our ability to study and manage radical-induced damage.

Antioxidant Defense Mechanisms

Biological systems counteract radical damage through sophisticated antioxidant defenses, including enzymes like superoxide dismutase (SOD), catalase, and glutathione peroxidase, and small-molecule antioxidants like vitamins C and E and glutathione. SOD converts superoxide into less harmful species, while catalase decomposes hydrogen peroxide (H_2O_2) into water and oxygen, preventing accumulation of this potentially harmful oxidant. Glutathione neutralizes radicals and regenerates other antioxidants. This maintains cellular redox balance.

These defenses help prevent oxidative stress-related diseases. For instance, vitamin E protects lipid membranes by scavenging peroxyl radicals. Research has begun to explore synthetic antioxidants inspired by natural systems, such as mimetics of SOD, for therapeutic use. Challenges include how to deliver antioxidants to specific cellular sites and avoid disruption of beneficial radical signaling. Nanotechnology, like antioxidant-loaded liposomes, enhances targeted delivery, while personalized medicine tailors antioxidant therapies to individual needs.

Radical Signaling

Radicals like nitric oxide (•NO) serve as signaling molecules in biological systems. •NO is produced by nitric oxide synthase and regulates vasodilation, neurotransmission, and immune responses. For example, it relaxes blood vessels, which lowers blood pressure, and modulates synaptic plasticity in the brain. Other radicals, like reactive oxygen species (ROS), act as secondary messengers in cellular pathways and influence processes like cell growth and apoptosis.

The discovery of radical signaling has reshaped our knowledge of redox biology. However, maintaining the delicate balance between signaling and damaging roles of radicals is challenging. Excessive •NO, for instance, can react with superoxide to form peroxynitrite, a harmful species. Research is focused on how to design selective radical modulators for therapeutic applications, such as •NO donors for cardiovascular treatments. Advances in synthetic chemistry have produced molecules that mimic or regulate radical signaling, which opens new avenues in precision medicine.

Enzyme Mechanisms

Certain enzymes, particularly those in the radical S-adenosylmethionine (SAM) superfamily, use radical intermediates to catalyze complex reactions. These enzymes generate radicals through SAM cleavage and enable transformations like DNA repair, antibiotic biosynthesis, and vitamin production. For example, the enzyme lysine 2,3-aminomutase catalyzes the rearrangement of lysine using radical intermediates, a process that is vital for bacterial metabolism.

Radical-based enzyme mechanisms are highly efficient and selective, inspiring synthetic catalysts. Their study has led to applications in biotechnology, such as engineering microbes to produce biofuels or pharmaceuticals. Challenges include understanding the intricate mechanisms of radical enzymes so as to harness them for industrial use. Advances in structural biology, like cryo-electron microscopy, elucidate these mechanisms, while synthetic biology expands its applications in sustainable chemical production.

Immune Response

ROS, including superoxide and hydrogen peroxide (not hydroxyl radicals as a primary product), are produced by immune cells like neutrophils and macrophages during the oxidative burst, a process that kills pathogens. This radical-mediated defense is important to combat bacterial and fungal infections. For example, myeloperoxidase in neutrophils generates hypochlorous acid from ROS and enhances antimicrobial activity.

The oxidative burst is a powerful immune tool, but excessive ROS production can damage host tissues, which contributes to inflammation-related diseases. Research is exploring ways to modulate ROS production for therapeutic purposes, such as how to boost immunity or reduce inflammation. Challenges include balancing antimicrobial efficacy with tissue safety. Nanotechnology, like ROS-responsive drug delivery systems and synthetic antioxidants, continues to be developed to fine-tune immune responses. This improves outcomes in infectious and inflammatory diseases.

Chapter 9: Aromatic Compounds

Aromatic compounds are characterized by their unique stability and reactivity. They are a cornerstone of organic chemistry. This stability arises from the delocalization of π electrons within a cyclic, planar structure, usually following Hückel's rule ($4n+2$ π electrons). Benzene, the archetypal aromatic compound, exemplifies this stability and serves as the foundation for a vast array of aromatic derivatives. Although aromatic rings are inherently stable, they are also susceptible to a variety of chemical transformations. However, unlike alkenes, which readily undergo addition reactions, aromatic compounds usually undergo substitution reactions. This preference for substitution preserves the aromaticity of the ring, maintaining its inherent stability.

Electrophilic Aromatic Substitution (EAS)

EAS is the most characteristic reaction of aromatic compounds. In this process, an electrophile (E^+), an electron-seeking species, attacks the electron-rich π system of the aromatic ring, which results in the replacement of a hydrogen atom with the electrophile. The reaction proceeds through a two-step mechanism that involves the formation of a resonance-stabilized intermediate.

Mechanism of Electrophilic Aromatic Substitution

The EAS mechanism consists of two key steps: electrophilic attack and deprotonation.

1) Electrophilic Attack and Formation of the Arenium Ion (σ-Complex)

The electrophile (E^+) is attracted to the delocalized π electrons of the aromatic ring. It forms a sigma (σ) bond with one of the carbon atoms, disrupts the aromaticity and creates a positively charged intermediate known as the arenium ion, sigma complex, or Wheland intermediate. This intermediate is resonance-stabilized, with the positive charge delocalized across several carbon atoms of the ring. The formation of this intermediate is the rate-determining step of the reaction.

The arenium ion is not aromatic because the sp^3 hybridization of the carbon bonded to the electrophile disrupts the continuous π system. However, the positive charge is delocalized through resonance, which makes the arenium ion more stable than a non-stabilized carbocation.

2) Deprotonation and Regeneration of Aromaticity

The arenium ion, while resonance-stabilized, has lost its aromaticity. In the second, faster step, a base (usually a conjugate base of the acid catalyst used to generate the electrophile) abstracts a proton from the carbon atom that is now bonded to the electrophile. This deprotonation restores the aromatic π system, yields the substituted aromatic product, and regenerates a proton. The deprotonation step is usually fast because the driving force is the restoration of aromaticity, which is thermodynamically favorable.

Key Electrophilic Aromatic Substitution Reactions

Several important reactions fall under the umbrella of EAS. These reactions involve the introduction of various functional groups onto the aromatic ring.

Nitration

Nitration is the introduction of a nitro group ($-NO_2$) onto the aromatic ring. This is usually achieved with a mixture of concentrated nitric acid (HNO_3) and sulfuric acid (H_2SO_4). Sulfuric acid protonates nitric acid and facilitates the formation of the nitronium ion (NO_2^+), the active electrophile; it acts as both a proton donor and dehydrating agent, not a true catalyst, since it is consumed in intermediate steps.

The mechanism of nitration involves the following steps:

1) Formation of the Nitronium Ion: Sulfuric acid protonates nitric acid and forms a protonated nitric acid species. This species then loses water to generate the nitronium ion (NO_2^+), the active electrophile.

- $HNO_3 + H_2SO_4 \rightleftharpoons H_2NO_3^+ + HSO_4^-$
- $H_2NO_3^+ \rightarrow NO_2^+ + H_2O$.

2) Electrophilic Attack: NO_2^+attacks the aromatic ring and forms the arenium ion intermediate.

3) Deprotonation: A base (usually HSO_4^-) removes a proton from the arenium ion, which restores aromaticity and yields nitrobenzene.

Nitrobenzene is a key intermediate in the synthesis of various aromatic compounds, particularly amines, through reduction (e.g., using Sn/HCl or H_2/Pd).

Halogenation (Chlorination and Bromination)

Halogenation is the substitution of a hydrogen atom with a halogen (chlorine or bromine). This reaction requires the presence of a Lewis acid catalyst such as ferric chloride ($FeCl_3$) or ferric bromide ($FeBr_3$), respectively. The Lewis acid polarizes the halogen molecule and makes it a stronger electrophile.

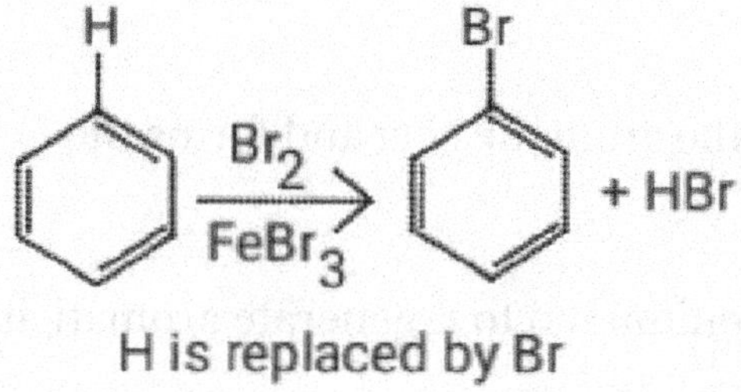

H Br

Br_2 / $FeBr_3$ + HBr

H is replaced by Br

Fig 31: Halogenation.

The mechanism of halogenation involves the following steps:

1) Activation of the Halogen: The Lewis acid catalyst (e.g., $FeCl_3$) interacts with the halogen molecule (e.g., Cl_2). This polarizes the halogen-halogen bond and generates a more electrophilic species.

- $Cl_2 + FeCl_3 \rightleftharpoons Cl^+ + FeCl_4^-$

2) Electrophilic Attack: The polarized halogen molecule attacks the aromatic ring and forms the arenium ion intermediate.

3) Deprotonation: A base (usually $FeCl_4^-$) removes a proton from the arenium ion, which restores aromaticity and yields the halobenzene.

Halobenzenes are important intermediates in various synthetic transformations, including nucleophilic aromatic substitution and Grignard reagent formation. Fluorination and iodination are less common and require specialized reagents. Fluorine is too reactive and can lead to uncontrolled reactions, while iodine is not reactive enough and requires strong oxidizing agents.

Sulfonation

Sulfonation is the introduction of a sulfonic acid group ($-SO_3H$) onto the aromatic ring. This usually involves benzene heated with fuming sulfuric acid (H_2SO_4 that contains dissolved SO_3) or concentrated sulfuric acid at elevated temperatures. The electrophile is sulfur trioxide (SO_3).

The mechanism of sulfonation involves the following steps:

1) Formation of the Electrophile: Sulfur trioxide (SO_3) is the active electrophile. It can be formed from the equilibrium of sulfuric acid:

- $2H_2SO_4 \rightleftharpoons SO_3 + H_3O^+ + HSO_4^-$

2) Electrophilic Attack: Sulfur trioxide attacks the aromatic ring and forms the arenium ion intermediate.

3) Proton Transfer: A proton is lost from the arenium ion to regenerate aromaticity, forming benzenesulfonic acid.

Benzenesulfonic acid and its derivatives are important in the detergent industry and as intermediates in the synthesis of other aromatic compounds (e.g., phenols through alkali fusion). Sulfonation is also reversible under acidic conditions. Heating benzenesulfonic acid in dilute sulfuric acid reverses the reaction to benzene.

Friedel-Crafts Alkylation

Friedel-Crafts alkylation is the introduction of an alkyl group onto the aromatic ring using an alkyl halide (R-X) in the presence of a Lewis acid catalyst, most commonly aluminum chloride ($AlCl_3$). The Lewis acid helps generate a carbocation or polarizes the alkyl halide, which makes the carbon atom electrophilic.

The mechanism of Friedel-Crafts alkylation involves the following steps:

1) Formation of the Electrophile: The Lewis acid catalyst (e.g., $AlCl_3$) interacts with the alkyl halide (R-X) and forms a carbocation or a polarized complex.

- $R\text{-}X + AlCl_3 \rightleftharpoons R^+ [AlCl_4]^-$ (or $R\delta^+$---$X\delta^-$---$AlCl_3$)

2) Electrophilic Attack: The carbocation or polarized complex attacks the aromatic ring and forms the arenium ion intermediate.

3) Deprotonation: A base (usually $AlCl_4^-$) removes a proton from the arenium ion, which restores aromaticity and yields the alkylated benzene.

Limitations of Friedel-Crafts Alkylation

- **Carbocation Rearrangements:** If the alkyl halide can form a primary or secondary carbocation, it may rearrange to a more stable carbocation (secondary or tertiary) through 1,2-hydride or alkyl shifts, which results in a mixture of alkylated products. This is a significant limitation, as it can lead to the formation of undesired isomers.
- **Polyalkylation:** The introduction of an electron-donating alkyl group activates the aromatic ring toward further electrophilic attack, which often results in polyalkylated products. Alkyl groups are activating groups and make the alkylated product more reactive than the starting material.
- **Reaction Failure with Deactivated Rings:** Friedel-Crafts alkylation does not occur with aromatic rings that bear strong EWGs, as they deactivate the ring toward electrophilic attack and can complex with the Lewis acid catalyst. Strongly EWGs reduce the electron density of the aromatic ring and make it less susceptible to electrophilic attack.
- **Intramolecular Alkylation:** Dihalides can undergo intramolecular Friedel-Crafts alkylation to form cyclic compounds and provide a route to cyclic systems. This is a useful synthetic strategy for forming rings.

Friedel-Crafts Acylation

Friedel-Crafts acylation is the introduction of an acyl group (RCO-) onto the aromatic ring using an acyl chloride (RCOCl) or a carboxylic acid anhydride—$(RCO)_2O$)—in the presence of a Lewis acid catalyst, usually aluminum chloride ($AlCl_3$). The electrophile is an acylium ion (RCO^+), which is resonance-stabilized.

The mechanism of Friedel-Crafts acylation involves the following steps:

1) Formation of the Electrophile: The Lewis acid catalyst (e.g., $AlCl_3$) interacts with the acyl chloride (RCOCl) and forms the acylium ion.

- $RCOCl + AlCl_3 \rightarrow R{-}C^{+}{\equiv}O + AlCl_4^-$

2) Electrophilic Attack: The acylium ion attacks the aromatic ring and forms the arenium ion intermediate.

3) Deprotonation: A base (usually $AlCl_4^-$) removes a proton from the arenium ion, which restores aromaticity and yields the acylated benzene.

Advantages of Friedel-Crafts Acylation over Alkylation

- **No Carbocation Rearrangements:** Acylium ions are resonance-stabilized and do not undergo rearrangements. This is a major advantage over Friedel-Crafts alkylation.
- **No Polyacylation:** The introduction of an electron-withdrawing acyl group deactivates the aromatic ring toward further electrophilic attack and makes polyacylation less likely. Acyl groups are deactivating groups and reduce the reactivity of the acylated product.

The resulting aromatic ketone can be subsequently reduced (e.g., through Clemmensen or Wolff-Kishner reduction) to yield an alkylated benzene without the rearrangement issues associated with direct Friedel-Crafts alkylation. This provides a workaround to the limitations of Friedel-Crafts alkylation.

Activating and Deactivating Groups

Substituents already present on an aromatic ring exert a significant influence on the rate and regioselectivity of subsequent EAS reactions. These substituents are classified based on their effect on the electron density of the aromatic ring and their ability to stabilize the arenium ion intermediate.

Activating Groups

Activating groups increase the electron density of the aromatic ring, which makes it more nucleophilic and thus more reactive toward electrophiles compared to benzene. They achieve this through electron donation through resonance (+M or +R effect) or inductive effects (+I effect).

- **Resonance Effects (+M or +R):** Activating groups with lone pairs of electrons directly attached to the ring can donate electron density through resonance, which significantly increases electron density, especially at the ortho and para positions.
- **Inductive Effects (+I):** Alkyl groups donate electron density through inductive effects, although this effect is weaker than resonance donation.

Strongly Activating Groups: These groups possess lone pairs of electrons directly attached to the ring that can be readily donated through resonance, significantly increasing electron density, especially at the ortho and para positions. Examples include:

- Amino group ($-NH_2$).
- Substituted amino groups ($-NHR$, $-NR_2$).
- Hydroxyl group (-OH).
- Alkoxy groups (-OR).

Moderately Activating Groups: These groups also donate electrons through resonance, but the lone pairs are involved in resonance elsewhere, which makes their donation to the ring less effective. Examples include:

- Amides (-NHCOR).
- Esters (-OCOR).

Weakly Activating Groups: These groups mainly donate electrons through hyperconjugation (overlap of σ bonds with the π system) and inductive effects (+I) due to the electron-releasing nature of alkyl groups. Examples include:

- Alkyl groups ($-CH_3$, -R).

Deactivating Groups

Deactivating groups decrease the electron density of the aromatic ring, which makes it less nucleophilic and thus less reactive toward electrophiles compared to benzene. They achieve this through electron withdrawal through resonance (-M or -R effect) or inductive effects (-I effect).

- **Resonance Effects (-M or -R):** Deactivating groups with π systems directly attached to the ring can withdraw electron density through resonance, which decreases electron density, especially at the ortho and para positions.
- **Inductive Effects (-I):** Electronegative atoms or groups withdraw electron density through inductive effects.

Strongly Deactivating Groups: These groups strongly withdraw electron density from the ring through resonance and/or strong inductive effects. Examples include:

- Nitro group ($-NO_2$).
- Cyano group (-CN).
- Trifluoromethyl group ($-CF_3$).

- Quaternary ammonium groups ($-NR_3^+$).

Moderately Deactivating Groups: These groups withdraw electron density through resonance that involves a carbonyl group directly attached to the ring. The electronegative oxygen atom pulls electron density away from the ring. Examples include:

- Carboxylic acids (-COOH).
- Esters (-COOR).
- Amides ($-CONH_2$).
- Aldehydes (-CHO).
- Ketones (-COR).

Weakly Deactivating Groups: Halogens (-F, -Cl, -Br, -I) are unique. They are deactivating due to their electronegativity (-I effect), which withdraws electron density from the ring. However, they also possess lone pairs that can participate in resonance (+M effect), although this effect is weaker than their inductive withdrawal.

Ortho/Meta/Para Directing Effects

Substituents on the aromatic ring not only influence the rate of EAS but also the position at which the incoming electrophile will preferentially attack. This is known as the directing effect, and substituents are classified as ortho/para directing or meta directing.

Ortho/Para Directing Groups

These groups direct the incoming electrophile to the ortho (positions 2 and 6 relative to the substituent) and para (position 4 relative to the substituent) positions.

- **Activating Ortho/Para Directors:** All activating groups (strong, moderate, and weak) are ortho/para directing. This is because the resonance stabilization of the arenium ion intermediate is most effective when the electrophile attacks the ortho or para positions. The lone pairs or σ bonds of these groups can participate in resonance structures that delocalize the positive charge onto the carbon atom that bears the activating substituent. This provides extra stability to the intermediate formed through ortho or para attack.

For example, in phenol (benzene with an -OH group), electrophilic attack at the ortho or para position allows for resonance structures where the positive charge is adjacent to the oxygen atom, and the lone pairs on oxygen can stabilize this positive charge.

- **Deactivating Ortho/Para Directors:** Halogens are the only deactivating groups that are ortho/para directing. Although their electronegativity (-I effect) withdraws electron density from the ring overall and makes it less reactive, their lone pairs can still participate in resonance (+M effect). When an electrophilic attack occurs at the ortho or para position, one of the resonance structures of the arenium ion places a positive charge adjacent to the halogen and allows for stabilization through the donation of a lone pair from the halogen. This resonance stabilization is only possible when the electrophile attacks the ortho or para positions; no such stabilization occurs for meta attack, which is why halogens are ortho/para directors despite being deactivating overall.

Meta Directing Groups

These groups direct the incoming electrophile to the meta position (positions three and five relative to the substituent).

All meta directing groups are deactivating. These groups withdraw electron density from the aromatic ring through inductive (-I) and/or resonance (-M) effects, reducing electron density especially at the ortho and para positions, thereby destabilizing arenium ion intermediates formed by electrophilic attack at those positions. When an electrophilic attack occurs at the ortho or para position on a benzene ring that bears a meta-directing group, one of the resonance structures of the arenium ion places the positive charge directly on the carbon atom bonded to the electron-withdrawing group. This is energetically unfavorable as the EWG destabilizes the positive charge. Attack at the meta position avoids this destabilizing resonance structure, which makes it the preferred site for electrophilic attack.

For example, in nitrobenzene (benzene with a -NO_2 group), the nitro group withdraws electron density through resonance. Ortho or para attack causes resonance structures where the positive charge is placed on the carbon bonded to the strongly electron-withdrawing nitro group, which destabilizes the intermediate due to the nitro group's inability to stabilize a neighboring positive charge. Meta attack avoids this unfavorable situation.

Exceptions and Considerations

- **Steric Effects:** Bulky substituents can hinder substitution at the ortho positions due to steric hindrance, even if they are ortho/para directing. This can lead to a preference for para-substitution.
- **Electronic Effects vs. Steric Effects:** Sometimes, a balance between electronic and steric factors determines the product distribution.
- **Multiple Substituents:** When multiple substituents are present on the aromatic ring, the directing effects can be additive or conflicting. Generally, strongly activating groups exert a stronger directing influence. If the directing effects conflict, the more strongly activating group often determines the major product, or steric factors may play a significant role.

Nucleophilic Aromatic Substitution (NAS)

NAS involves the displacement of a leaving group on an aromatic ring by a nucleophile. Unlike EAS, which occurs on electron-rich rings, NAS usually requires the aromatic ring to be electron-deficient, usually due to the presence of strong EWGs, especially at ortho and para positions to the leaving group.

There are two main mechanisms for NAS:

SNAr (Addition-Elimination) Mechanism

This mechanism is favored when the aromatic ring has strong EWGs (e.g., nitro groups) ortho or para to the leaving group (usually a halide).

The mechanism involves two steps:

1) Nucleophilic Attack and Formation of the Meisenheimer Complex (σ-Complex): The nucleophile attacks the carbon atom that bears the leaving group and forms a resonance-stabilized carbanion intermediate known as the Meisenheimer complex or σ-complex. The EWGs stabilize the negative charge in this intermediate through resonance delocalization.

2) Elimination of the Leaving Group: The Meisenheimer complex is unstable due to the negative charge and loss of aromaticity. It eliminates the leaving group to regenerate the aromatic system with the nucleophile now substituted in place of the leaving group.

Factors Favoring SNAr Mechanism

- **Strong EWGs:** Presence of $-NO_2$, -CN, etc., especially at ortho and para positions to the leaving group, delocalizes the negative charge and stabilizes the Meisenheimer complex. The more EWGs and the closer they are to the reaction center (ortho/para), the faster the reaction.
  Good Leaving Group: The leaving group must be stable once it departs with a negative charge. Common leaving groups include halides (e.g., Cl^-, Br^-, F^-); the nitro group ($-NO_2$) is not a leaving group but an activating group that stabilizes the Meisenheimer complex. Fluoride is often a better leaving group than other halides in polar aprotic solvents due to the high electronegativity of fluorine that enhances the electrophilicity of the carbon it's attached to, although the trend can vary with solvent and reaction conditions.
  Strong Nucleophile: A good nucleophile with a significant negative charge or a lone pair of electrons is necessary to initiate the attack. Examples include hydroxide (OH^-), alkoxides (RO^-), amines (RNH_2), and carbanions.

Benzyne Mechanism (Elimination-Addition)

This mechanism occurs under harsh conditions (strong base, high temperature) and does not require activating groups on the aromatic ring. It involves the formation of a highly reactive intermediate called benzyne, which contains a strained triple bond within the six-membered ring.

The mechanism involves two steps:

1) Elimination to Form Benzyne: A strong base (e.g., sodium amide ($NaNH_2$)) abstracts a proton from a carbon atom adjacent to the carbon that bears the leaving group (usually a halide); potassium hydroxide (KOH) is generally not strong enough to generate benzyne under standard conditions. Simultaneously or subsequently, the leaving group departs, which forms the benzyne intermediate. This elimination is a β-elimination, similar to E2 reactions in aliphatic systems.

2) Nucleophilic Addition to Benzyne: The highly reactive benzyne intermediate, with its electron-deficient triple bond, is readily attacked by a nucleophile. The nucleophile can add to either of the triple-bonded carbons, and if the benzyne is unsymmetrical, this can lead to a mixture of products. The resulting carbanion intermediate is then protonated by the solvent or another proton source in the reaction mixture to yield the substituted aromatic product.

Characteristics of the Benzyne Mechanism

- **Requires a Strong Base:** To abstract a proton with a low pKa.
- **No Need for Activating Groups:** The reaction proceeds through a highly reactive intermediate rather than a direct nucleophilic attack on the aromatic ring.
- **Potential for Isomer Formation:** If the benzyne intermediate is unsymmetrical, the nucleophile can attack at either carbon of the triple bond, which results in a mixture of regioisomers. The position of the leaving group in the starting material may not be the same as the position of the nucleophile in the product.
- **Evidence for Benzyne Intermediate:** Trapping experiments with dienes (e.g., Diels-Alder reactions) can provide evidence for the formation of a benzyne intermediate.

Conjugated Systems and Aromaticity

Conjugated systems and aromaticity build upon previous discussions and explore the delocalization of electrons that results in unique stabilities and reactivities. This section will delve into conjugation, molecular orbital theory for dienes, the Diels-Alder reaction, and extended conjugated systems.

Conjugation and Stability

Conjugation refers to the overlap of p orbitals across three or more adjacent atoms in a molecule. This usually occurs when single and multiple bonds alternate (e.g., –C=C–C=C–) but can also involve lone pairs or a single p orbital on an adjacent atom (e.g., in allylic carbocations or radicals).

Key Aspects of Conjugation:

- Delocalization of Electrons: Overlapping p orbitals create a continuous π system and allow electrons to be delocalized over the entire conjugated structure. This delocalization lowers electron energy and enhances the stability of the molecule or ion.
- Resonance: Conjugated systems can be depicted by multiple resonance structures that illustrate the delocalization of π electrons. The actual structure is a resonance hybrid and represents a weighted average of these contributing structures.
- Bond Lengths: In conjugated systems, single bonds adjacent to multiple bonds are usually shorter and stronger than usual single bonds, while multiple bonds

are longer and weaker than isolated ones due to the partial double-bond character from resonance.

- UV-Vis Spectroscopy: Conjugated systems absorb UV and visible light at longer wavelengths and with greater intensity compared to isolated double bonds. Greater conjugation results in a smaller energy gap between the highest occupied molecular orbital (HOMO) and the lowest unoccupied molecular orbital (LUMO). This allows for the absorption of lower-energy light. This property contributes to the colors of many organic compounds.

Stability Enhancement from Conjugation

The delocalization of electrons in conjugated systems increases stability compared to non-conjugated counterparts, a phenomenon known as resonance energy or delocalization energy. For instance, 1,3-butadiene is more stable than would be predicted for two isolated double bonds. To estimate resonance energy, compare the experimental heat of hydrogenation of the conjugated system with that of the corresponding number of isolated double bonds.

Examples of Conjugated Systems with Enhanced Stability

- Allylic Carbocations and Radicals: The positive charge or unpaired electron is delocalized over three carbon atoms, which results in resonance stabilization. $CH_2=CH-CH_2^+$ ↔ $CH_2^+-CH=CH_2$ ↔ $CH_2=CH^+-CH_2$
- Aromatic Compounds: These cyclic, planar, conjugated systems contain a specific number of π electrons (4n + 2, according to Hückel's rule) and exhibit remarkable stability due to extensive cyclic delocalization. Benzene serves as a prime example.

Molecular Orbital (MO) Theory for Dienes

MO theory provides a more comprehensive view of bonding and electron delocalization in conjugated systems compared to resonance theory. In MO theory, atomic orbitals combine to create MOs that extend throughout the entire molecule.

1,3-Butadiene: Consider 1,3-butadiene ($CH_2=CH-CH=CH_2$), a simple conjugated diene with four sp^2 hybridized carbon atoms. Each carbon atom possesses one p orbital perpendicular to the molecular plane, which overlaps to form a π system.

- **Atomic Orbitals:** Four 2p atomic orbitals combine to produce four π MOs (π_1, π_2, π_3^*, π_4^*).

- **Energy Levels:** These four π MOs have varied energy levels, and the number of nodes (regions of zero electron density) increases with energy.

  1) π1: Lowest energy, no internal nodes (all p orbitals overlap in phase). This is a bonding MO.

  2) π2: One node (one antibonding interaction). Also a bonding MO but at a higher energy than π1.

  *3) π3:** Two nodes, one bonding and one antibonding interaction; this is a partially antibonding MO (LUMO in 1,3-butadiene).

  *4) π4 :** Three nodes with alternating phases; this is the highest-energy fully antibonding MO.

- Electron Occupancy: 1,3-butadiene contains four π electrons (one from each p orbital). These electrons occupy the two bonding π MOs (π1 and π2) with opposite spins and stabilize the molecule.

Other Conjugated Systems

MO theory principles can be applied to other conjugated systems with greater π electron counts, such as longer polyenes and cyclic conjugated systems (annulenes). For a linear conjugated system with n p orbitals, there will be n π MO.

Diels-Alder Reaction

The Diels-Alder reaction is a [4+2] cycloaddition process between a conjugated diene and a dienophile (a molecule with a double or triple bond). It results in a six-membered ring (a cyclohexene derivative) with high stereospecificity and regioselectivity. This reaction is widely used in organic synthesis for constructing cyclic systems.

Mechanism

The Diels-Alder reaction is a concerted, single-step process that proceeds through a cyclic transition state. The π electrons from the diene and the dienophile rearrange to form two new σ bonds and one new π bond in the cyclic product.

- Concerted Mechanism: All bond breaking and forming occur simultaneously in a single cyclic transition state, with no intermediates formed.

- Cyclic Transition State: The transition state features a cyclic arrangement of six π electrons (four from the diene and two from the dienophile) that interact cooperatively.

Stereospecificity

The Diels-Alder reaction is stereospecific for both the diene and the dienophile.

- Diene: The configuration of substituents on the diene is preserved in the product, but only if the diene is in the reactive s-cis conformation; syn addition refers specifically to the concerted nature of the bond formation, not necessarily to cis/trans geometry preservation on acyclic dienes. If the diene is in the s-cis conformation (essential for the reaction), the relative positions of the substituents are maintained in the cyclic product.
- Dienophile: The stereochemistry of substituents on the dienophile is also retained in the final product (syn addition).

Regioselectivity

When substituted dienes and dienophiles are used, the reaction often displays high regioselectivity and favors one constitutional isomer over others. To predict regiochemistry, consider the electronic effects of substituents on both the diene and the dienophile (e.g., align the largest coefficients of the HOMO of one reactant with the largest coefficients of the LUMO of the other).

Factors that Affect the Diels-Alder Reaction

- Diene Conformation: The diene must be in the s-cis conformation (where the two double bonds are adjacent) to participate in the reaction. Rotation around the single bond allows for interconversion between s-cis and s-trans conformations, but the s-cis conformation is often higher in energy due to steric interactions. Cyclic dienes that are locked in the s-cis form (e.g., cyclopentadiene) are particularly reactive in Diels-Alder reactions.
- Electron-Donating and EWGs: Dienes with EDGs usually react more rapidly with dienophiles that have EWGs. EDGs increase the energy of the diene's HOMO, while EWGs lower the LUMO energy of the dienophile, reduce the HOMO-LUMO gap and facilitate the reaction. Conversely, dienes with EWGs generally react more slowly, and such combinations with dienophiles bearing EDGs are less favorable due to an increased HOMO-LUMO gap, making the reaction less efficient.

- Steric Effects: Bulky substituents on the diene or dienophile can hinder the approach of reactants and slow the reaction.
- Catalysis: Lewis acids (e.g., $AlCl_3$, BF_3) can catalyze Diels-Alder reactions by coordinating with the dienophile (if it has a carbonyl or other Lewis basic group). This enhances its electrophilicity and reactivity toward the diene.

Applications of the Diels-Alder Reaction

The Diels-Alder reaction is a powerful tool in organic synthesis to:

- Construct complex cyclic molecules with defined stereochemistry and regiochemistry.
- Synthesize natural products, pharmaceuticals, and polymers.
- Perform tandem reactions (e.g., sequential Diels-Alder reactions) to build intricate structures in a single step.
- Create stereocenters and control the relative orientations of substituents in the cyclic product.

The predictable stereochemistry and regiochemistry, along with the variety of dienes and dienophiles that can participate, make the Diels-Alder reaction one of the most important and versatile reactions in organic chemistry.

Extended Conjugated Systems

Extended conjugated systems consist of more than two double bonds or a combination of double and triple bonds. This allows for greater delocalization of π electrons. Examples include polyenes (linear chains with alternating single and double bonds, such as β-carotene), polyynes (chains with alternating single and triple bonds), and aromatic systems with multiple rings (e.g., naphthalene, anthracene).

Properties of Extended Conjugated Systems

- Enhanced Stability: Increased delocalization of π electrons results in higher resonance energy and overall stability.
- Longer Wavelength UV-Vis Absorption: As conjugation extends, the HOMO-LUMO energy gap decreases and allows for the absorption of light at longer wavelengths. This is why many extended conjugated systems exhibit color. For instance, β-carotene absorbs blue and green light, which appears orange due to its long conjugated polyene chain.
- Unique Reactivities: Extended conjugated systems can undergo various reactions, including electrophilic substitution (especially in aromatic systems),

cycloaddition (like Diels-Alder in some cases), and radical reactions, with regioselectivity influenced by the delocalized π system.

- Conductivity in Polymers: Polymers with extended conjugated π systems in their backbone (e.g., polyacetylene, polypyrrole) can exhibit electrical conductivity. The delocalized electrons can move along the conjugated chain, which renders these materials “organic metals” or semiconductors, with applications in organic electronics.

Examples of Extended Conjugated Systems

A) Polyenes:

- β-Carotene: A tetraterpene with a long conjugated polyene chain, responsible for the orange color of carrots and a precursor to vitamin A.
- Retinal: A polyene aldehyde that serves as the light-absorbing molecule in the rhodopsin protein in the retina, essential for vision. The cis-trans isomerization of the retinal upon light absorption triggers a nerve impulse.

B) Polyynes: Linear chains of carbon atoms linked by alternating single and triple bonds. These are generally less stable than polyenes but possess interesting electronic properties.

C) Polycyclic Aromatic Hydrocarbons (PAHs): Fused benzene rings, such as naphthalene (two rings), anthracene (three linear rings), and phenanthrene (three angular rings). These compounds demonstrate aromaticity and distinct chemical and physical properties, with larger PAHs having potential carcinogenic effects.

D) Porphyrins and Related Compounds: Large cyclic conjugated systems that contain heteroatoms (like nitrogen). Heme (in hemoglobin) and chlorophyll (in plants) are biologically significant porphyrin derivatives with extended conjugation that play vital roles in oxygen transport and photosynthesis, respectively.

UV-Vis Spectroscopy of Conjugated Systems

UV-Vis spectroscopy is a highly effective analytical method employed to investigate the electronic structure of molecules, especially those with conjugated pi systems. Conjugation refers to the arrangement of alternating double and single bonds or the continuous overlap of p orbitals, which reduces the energy gap between MOs. This characteristic allows molecules to absorb light in the ultraviolet (UV) and visible (Vis) parts of the electromagnetic spectrum. The absorption of light causes electronic transitions, mainly from bonding π orbitals (HOMO) to antibonding π* orbitals

(LUMO). This process not only yields distinctive UV-Vis spectra but also often imparts color to the molecules being studied.

Electronic Transitions in Conjugated Systems

In conjugated molecules, the main electronic transition observed in UV-Vis spectroscopy is the $\pi \rightarrow \pi^*$ transition. This transition involves exciting an electron from the HOMO, a bonding π orbital, to the LUMO, an antibonding π^* orbital. The energy difference (ΔE) between these orbitals is vital, as it determines the wavelength (λ) of light absorbed, per the equation $E=hc/\lambda$, where h is Planck's constant and c is the speed of light.

For instance, isolated double bonds, like those in ethene, absorb light at shorter wavelengths (approximately 165–174 nm), which corresponds to higher energy transitions. However, conjugation significantly lowers the HOMO-LUMO energy gap, which results in a shift of the absorption maxima (λ_{max}) to longer wavelengths, a phenomenon known as a bathochromic shift. This shift often brings the absorption into the UV-visible range (220–700 nm) and makes it accessible for analysis with standard spectrophotometers.

Effect of Conjugation on UV-Vis Absorption

The degree of conjugation in a molecule has a direct impact on its absorption spectrum. Greater delocalization of π electrons occurs, which stabilizes the HOMO and destabilizes the LUMO, effectively narrowing the energy gap between them. This stabilization results in a shift of λ_{max} to longer wavelengths, which indicates lower energy transitions, and can even extend into the visible region and impart color to the molecule.

For example, 1,3-butadiene, the simplest conjugated diene, absorbs UV light at approximately 217 nm due to its conjugated system, while isolated double bonds absorb at shorter wavelengths.

Spectral Characteristics of Conjugated Systems

Several key spectral characteristics are associated with conjugated systems:

- Bathochromic Shift (Red Shift): As conjugation increases, the absorption maximum shifts to longer wavelengths. This indicates a decrease in energy requirements for electronic transitions.

- Hyperchromic Effect: The intensity of absorption, quantified by molar absorptivity (ε), increases with greater conjugation. This reflects a higher likelihood of electronic transitions.
- Multiple Absorption Peaks: Extensive conjugation can lead to multiple absorption bands or shoulders in the spectrum, which arise from different electronic transitions occurring within the system.

Practical Implications and Applications

The implications of UV-Vis spectroscopy extend to various practical applications.

- Color and Pigments: Natural pigments and dyes derive their colors from extensive conjugation, which shifts absorption into the visible spectrum. This makes them visually distinctive.
- Molecular Identification: UV-Vis spectroscopy is instrumental in identifying chromophores (the parts of molecules responsible for light absorption) and assessing the extent of conjugation in organic and biological molecules.
- Quantitative Analysis: This technique allows for the quantification of concentration and electronic properties because it measures absorbance at λ_{max}. This facilitates accurate analytical assessments.

From an MO perspective, conjugation causes the formation of multiple π MOs through the linear combination of adjacent p orbitals. In the case of 1,3-butadiene, four π MOs are formed, with the HOMO and LUMO being closer in energy compared to isolated double bonds. The reduced HOMO-LUMO gap corresponds to absorption at longer wavelengths, which further emphasizes the importance of conjugation when electronic properties and spectral characteristics are determined.

Chapter 10: Carbonyl Chemistry

Carbonyl compounds are compounds characterized by the presence of a carbonyl group (C=O). They are ubiquitous in organic chemistry and play a vital role in a wide range of chemical and biological processes. The carbonyl group consists of a carbon atom double-bonded to an oxygen atom. This seemingly simple functional group imparts a unique set of properties to the molecules that contain it. It makes them versatile building blocks for organic synthesis.

The carbon-oxygen double bond is highly polarized due to the greater electronegativity of oxygen compared to carbon. This polarization results in a partial positive charge (δ+) on the carbon atom and a partial negative charge (δ-) on the oxygen atom. This charge distribution makes the carbonyl carbon electrophilic—susceptible to attack by nucleophiles—and the carbonyl oxygen basic, meaning it can accept a proton or interact with electrophiles, but it is not typically nucleophilic.

Properties of Aldehydes and Ketones

Aldehydes and ketones are two closely related classes of carbonyl compounds that differ in their substitution pattern at the carbonyl carbon.

- Aldehydes: Have the general formula RCHO, where R is an alkyl or aryl group, and one substituent is always a hydrogen atom.
- Ketones: Have the general formula RCOR', where R and R' are alkyl or aryl groups.

This seemingly small difference in structure leads to significant differences in their physical and chemical properties.

Physical Properties

The physical properties of aldehydes and ketones are influenced by the polarity of the carbonyl group and the presence of other functional groups in the molecule.

- **Boiling Points:** Aldehydes and ketones have higher boiling points than alkanes of comparable molecular weight due to dipole-dipole interactions between the carbonyl groups. However, they have lower boiling points than alcohols of comparable molecular weight because they cannot form strong hydrogen bonds with themselves.
- **Solubility:** Lower molecular weight aldehydes and ketones are soluble in water due to their ability to form hydrogen bonds with water molecules. As the size of

the alkyl or aryl groups increases, the solubility in water decreases due to the increasing hydrophobic character of the molecule.
- **Odor:** Many aldehydes and ketones have characteristic odors. Some are pleasant (e.g., vanillin in vanilla extract, benzaldehyde in almonds), while others are pungent or unpleasant (e.g., formaldehyde).

Spectroscopic Properties

Spectroscopic techniques are valuable tools that help us identify and characterize aldehydes and ketones.

A) IR Spectroscopy: The carbonyl group exhibits a strong, characteristic absorption band in the IR spectrum typically around 1710–1740 cm^{-1} for aldehydes and ketones, with exact positions varying depending on conjugation and ring strain. The exact position of the band depends on the structure of the aldehyde or ketone and the presence of other functional groups. Aldehydes also show characteristic C-H stretching vibrations around 2700–2850 cm^{-1}.

B) NMR Spectroscopy:

- ^{1}H NMR: The aldehyde proton (CHO) appears as a distinct singlet in the region of δ 9–10 ppm, which is significantly downfield compared to other protons in the molecule.
- ^{13}C NMR: The carbonyl carbon (C=O) appears as a characteristic signal in the region of δ 190–220 ppm, which is also significantly downfield compared to other carbons in the molecule.

Nucleophilic Addition Reactions

The carbonyl group is highly susceptible to nucleophilic attack due to the electrophilic nature of the carbonyl carbon. Nucleophilic addition reactions are a hallmark of carbonyl chemistry. In these reactions, a nucleophile (Nu^-) attacks the carbonyl carbon, breaks the π bond, and forms a tetrahedral intermediate. The subsequent fate of this intermediate depends on the nature of the nucleophile and the reaction conditions.

General Mechanism of Nucleophilic Addition

The general mechanism of nucleophilic addition to a carbonyl group involves the following steps:

- Nucleophilic Attack: The nucleophile (Nu^-) attacks the electrophilic carbonyl carbon, forms a new σ bond, and breaks the π bond. This generates a tetrahedral intermediate with a negative charge on the oxygen atom.
- Protonation: The oxygen atom of the tetrahedral intermediate is protonated by an acid (H^+) to give a neutral addition product.

Specific Nucleophilic Addition Reactions

Several important reactions fall under the umbrella of nucleophilic addition to carbonyl compounds.

Hydration

Water can act as a nucleophile and add to the carbonyl group to form a gem-diol (a diol with both hydroxyl groups on the same carbon). This reaction is usually catalyzed by an acid or base.

- Acid-Catalyzed Hydration: The carbonyl oxygen is protonated, which makes the carbonyl carbon more electrophilic. Water then attacks the carbonyl carbon, followed by deprotonation to give the gem-diol.
- Base-Catalyzed Hydration: Hydroxide ion (OH^-) attacks the carbonyl carbon, followed by protonation to give the gem-diol.

Gem-diols are generally unstable and readily revert to the carbonyl compound when they lose water. However, in some cases, such as formaldehyde, the gem-diol is the predominant form in aqueous solution.

Acetal Formation

Aldehydes and ketones react with alcohols in the presence of an acid catalyst to form acetals (from aldehydes) or ketals (from ketones). This reaction involves two successive nucleophilic additions of alcohol molecules.

The mechanism of acetal formation involves the following steps:

1) Protonation of the Carbonyl Oxygen: The carbonyl oxygen is protonated by the acid catalyst, which makes the carbonyl carbon more electrophilic.

2) Nucleophilic Attack by Alcohol: An alcohol molecule attacks the carbonyl carbon and forms a hemiacetal (or hemiketal) intermediate.

3) Protonation of the Hydroxyl Group: The hydroxyl group of the hemiacetal is protonated, which converts it into a good leaving group (water).

4) Loss of Water: Water is eliminated and generates an oxonium ion.

5) Nucleophilic Attack by Alcohol: Another alcohol molecule attacks the oxonium ion and forms the acetal (or ketal).

6) Deprotonation: A proton is removed to give the neutral acetal (or ketal) product.

Acetals and ketals are stable under neutral and basic conditions but are readily hydrolyzed back to the carbonyl compound and alcohol under acidic conditions. Acetals and ketals are often used as protecting groups for aldehydes and ketones, as they can be easily introduced and removed under mild conditions.

Imine and Enamine Formation

Aldehydes and ketones react with primary amines (RNH_2) to form imines (also called Schiff bases). This reaction involves the nucleophilic addition of the amine to the carbonyl group, followed by the elimination of water.

The mechanism of imine formation involves the following steps:

- Nucleophilic Attack by Amine: The primary amine attacks the carbonyl carbon and forms a tetrahedral intermediate called a carbinolamine.
- Proton Transfer: A proton transfer occurs within the carbinolamine intermediate, typically from the nitrogen to a solvent or catalyst, and then from a solvent or catalyst to the hydroxyl group to facilitate water elimination.
- Elimination of Water: Water is eliminated, forming an iminium ion intermediate, which then loses a proton to give the imine.

Imines are important intermediates in many biological processes, such as the reactions catalyzed by pyridoxal phosphate (vitamin B6).

Aldehydes and ketones react with secondary amines (R_2NH) to form enamines. This reaction also involves the nucleophilic addition of the amine to the carbonyl group, followed by the elimination of water. However, in this case, the elimination of water occurs from a carbon atom adjacent to the carbonyl group, which results in the formation of a carbon-carbon double bond.

The mechanism of enamine formation involves the following steps:

- Nucleophilic Attack by Amine: The secondary amine attacks the carbonyl carbon, leading to formation of a carbinolamine intermediate after protonation.
- Proton Transfer: A proton transfer occurs within the carbinolamine intermediate, typically from the nitrogen to a solvent or catalyst, and then from a solvent or catalyst to the hydroxyl group to facilitate water elimination.
- Elimination of Water: Water is eliminated, which generates an iminium ion.
- Deprotonation: A proton is removed from a carbon atom adjacent to the carbonyl group, which generates the enamine.

Enamines are useful synthetic intermediates, as they can act as nucleophiles in various reactions.

Wittig Reaction

The Wittig reaction is a powerful method used to convert aldehydes and ketones into alkenes. This reaction involves the reaction of a carbonyl compound with a phosphorus ylide (also called a Wittig reagent).

A phosphorus ylide is a compound that contains a negatively charged carbon atom bonded to a positively charged phosphorus atom. Wittig reagents are usually prepared by reacting a triphenylphosphonium salt with a strong base.

The mechanism of the Wittig reaction involves the following steps:

- Nucleophilic Attack by Ylide: The carbon atom of the ylide attacks the electrophilic carbonyl carbon, forming a betaine intermediate (a zwitterionic species), although in many cases the reaction proceeds directly through a concerted [2+2] cycloaddition to form the oxaphosphetane without isolable betaine.
- Formation of Oxaphosphetane: The betaine intermediate cyclizes to form an oxaphosphetane intermediate.
- Elimination: The oxaphosphetane intermediate decomposes to form the alkene and triphenylphosphine oxide.

The Wittig reaction is a versatile method used to synthesize alkenes, with the stereochemistry (E or Z) influenced by the nature of the ylide; stabilized ylides often give E-alkenes, while non-stabilized ylides tend to favor Z-alkenes.

Grignard Reaction

Grignard reagents (RMgX, where X is a halogen) are organometallic compounds that are strong nucleophiles and strong bases. They react with aldehydes and ketones to form alcohols.

The mechanism of the Grignard reaction involves the following steps:

- Nucleophilic Attack by Grignard Reagent: The Grignard reagent attacks the carbonyl carbon and forms a tetrahedral intermediate.
- Protonation: The alkoxide intermediate is protonated by the addition of dilute acid to give the alcohol.

The Grignard reaction is a versatile method used to synthesize a wide variety of alcohols.

Addition of Hydrogen Cyanide (HCN)

Aldehydes and ketones react with hydrogen cyanide (HCN) to form cyanohydrins. This reaction is usually carried out using a cyanide salt (e.g., NaCN or KCN) in the presence of an acid catalyst.

The mechanism of cyanohydrin formation involves the following steps:

- Nucleophilic Attack by Cyanide: The cyanide ion (CN^-) attacks the carbonyl carbon and forms a tetrahedral intermediate.
- Protonation: The alkoxide intermediate is protonated by the acid catalyst to give the cyanohydrin.

Cyanohydrins are useful synthetic intermediates, as they can be converted into a variety of other functional groups, such as α-hydroxy acids and α-amino acids.

Carboxylic Acids and Derivatives

Carboxylic acids are organic compounds that contain a carboxyl group (-COOH). They are characterized by their acidity and their ability to form a variety of derivatives, such as esters, amides, acid halides, and anhydrides.

Properties of Carboxylic Acids

- Acidity: Carboxylic acids are weak acids, with pKa values usually in the range of 4–5. The acidity is due to the resonance stabilization of the carboxylate anion formed upon deprotonation.

- Hydrogen Bonding: Carboxylic acids can form strong hydrogen bonds with themselves and with other molecules. This leads to relatively high boiling points and good solubility in polar solvents.
- Spectroscopic Properties: Carboxylic acids exhibit characteristic IR absorptions for the O-H stretch (2500–3300 cm^{-1}) and the C=O stretch (1700–1725 cm^{-1}). The ^{1}H NMR spectrum shows a characteristic broad signal for the carboxyl proton (COOH) at δ 10–13 ppm.

Reactions of Carboxylic Acids

Carboxylic acids undergo a variety of reactions, including:

- Neutralization: Carboxylic acids react with bases to form carboxylate salts.
- Esterification: Carboxylic acids react with alcohols in the presence of an acid catalyst to form esters. This reaction is known as Fischer esterification.
- Amide Formation: Carboxylic acids react with amines to form amides, but this reaction typically requires activation of the carboxylic acid—commonly via conversion to an acid chloride or by using coupling agents—because direct condensation is not efficient under standard conditions.
- Reduction: Carboxylic acids can be reduced to primary alcohols using strong reducing agents such as lithium aluminum hydride ($LiAlH_4$).
- Decarboxylation: Carboxylic acids can undergo decarboxylation under certain conditions.

Carboxylic Acid Derivatives

Carboxylic acid derivatives are compounds in which the hydroxyl group of the carboxyl group has been replaced by another group. Common carboxylic acid derivatives include:

- Esters (RCOOR’): Formed by the reaction of a carboxylic acid with an alcohol.

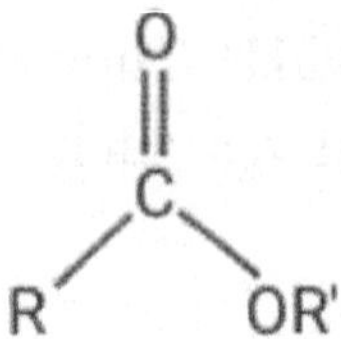

Fig 32: General structure of esters.

- Amides (RCONR’R”): Formed by the reaction of a carboxylic acid with an amine.

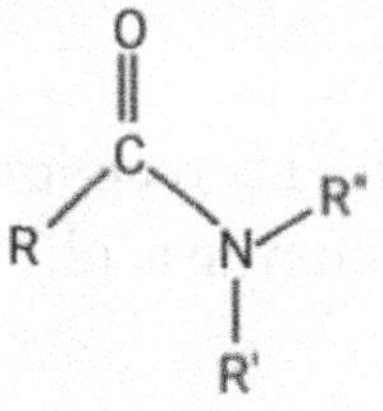

Fig 33: General structure of amides.

- Acid Halides (RCOX, where X = Cl, Br): Formed by the reaction of a carboxylic acid with a halogenating agent such as thionyl chloride ($SOCl_2$), phosphorus pentachloride (PCl_5), or oxalyl chloride ($(COCl)_2$).

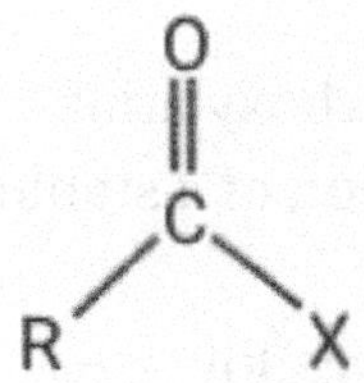

Fig 34: General structure of acid halides.

- Anhydrides (RCOOCOR'): Formed by the dehydration (loss of water) between two carboxylic acid molecules.

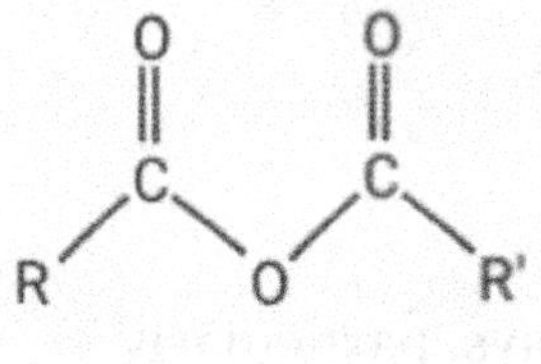

Fig 35: General structure of anhydrides.

These derivatives undergo nucleophilic acyl substitution reactions, where a nucleophile attacks the carbonyl carbon, and the leaving group was originally attached to the carbonyl carbon. The reactivity of carboxylic acid derivatives toward nucleophilic acyl substitution depends on the nature of the leaving group. Acid halides are the most reactive, followed by anhydrides, esters, and amides.

Decarboxylation and Related Transformations

Decarboxylation is the loss of carbon dioxide (CO_2) from a carboxylic acid. This reaction is often facilitated by the presence of specific structural features in the carboxylic acid, such as a β-keto group or a β-dicarboxylic acid.

Mechanism of Decarboxylation

The mechanism of decarboxylation usually involves a cyclic transition state in which the carboxyl group is lost as carbon dioxide, and a proton is transferred to the carbon atom adjacent to the carboxyl group.

Examples:

- Decarboxylation of β-Keto Acids: β-Keto acids readily undergo decarboxylation upon heating to form ketones. This reaction is driven by the formation of a stable enol intermediate, which then tautomerizes to the ketone.
- Decarboxylation of β-Dicarboxylic Acids: β-Dicarboxylic acids also readily undergo decarboxylation upon heating to form carboxylic acids.
- Hunsdiecker Reaction: The Hunsdiecker reaction is a method used to convert silver carboxylates (derived from carboxylic acids) to alkyl halides with the loss of carbon dioxide. This reaction involves the reaction of a silver carboxylate salt with a halogen (Cl_2 or Br_2).

Biological Significance of Decarboxylation

1) Role in Cellular Metabolism

Decarboxylation reactions are essential for various metabolic pathways, particularly those related to energy production and the transformation of biomolecules. These reactions usually involve specific enzymes, such as decarboxylases, which help remove CO_2 from carboxylic acids. Key metabolic processes that rely on decarboxylation include the following reactions.

A) Citric Acid Cycle (Krebs Cycle)

The citric acid cycle, a vital component of cellular respiration, depends on decarboxylation to convert isocitrate to α-ketoglutarate and then α-ketoglutarate to succinyl-CoA. Enzymes like isocitrate dehydrogenase and α-ketoglutarate dehydrogenase catalyze these reactions, release CO_2, and produce NADH, which is an important electron carrier. The NADH generated is utilized in the electron transport chain to produce ATP, which links carbohydrate, fat, and protein metabolism to energy production.

B) Fatty Acid β-Oxidation

In the β-oxidation of fatty acids, decarboxylation does not occur; instead, the process involves the sequential removal of two-carbon units as acetyl-CoA, which then undergoes decarboxylation during the citric acid cycle. This acetyl-CoA then enters the citric acid cycle, where additional decarboxylation produces CO_2 and energy. This pathway is important for lipid metabolism and provides an alternative energy source during fasting or prolonged exercise.

C) Amino Acid Metabolism

Decarboxylation is a vital step in the breakdown of certain amino acids. For instance:

- Histidine is converted to histamine, a neurotransmitter involved in immune responses and gastric acid secretion.
- Glutamate is transformed into γ-aminobutyric acid (GABA), an inhibitory neurotransmitter important for regulating neuronal excitability.
- Tryptophan is decarboxylated to form tryptamine, a precursor to serotonin, which influences mood, sleep, and appetite. These processes illustrate how decarboxylation generates biologically active molecules from amino acids.

2) Biosynthesis of Neurotransmitters and Signaling Molecules

Decarboxylation plays a key role in the synthesis of various neurotransmitters and signaling molecules essential for cellular communication. The enzymatic removal of CO_2 from amino acids or other precursors leads to the formation of amines or bioactive compounds. Examples include:

- Dopamine is produced from the decarboxylation of L-DOPA, while norepinephrine is synthesized from dopamine via hydroxylation, not decarboxylation. They are key catecholamines involved in motor control and stress responses.
- Serotonin (5-hydroxytryptamine) is synthesized from 5-hydroxytryptophan through decarboxylation. It affects mood and digestion.
- Polyamines (e.g., putrescine, spermidine, and spermine) are derived from the decarboxylation of ornithine. They are involved in cell growth and DNA stabilization.

These compounds highlight the importance of decarboxylation in the regulation of physiological processes, from neural signaling to immune function.

3) Regulation of pH and Acid-Base Balance

Decarboxylation reactions help regulate cellular and systemic pH because they produce CO_2, which can be converted to bicarbonate (HCO_3^-) through the action of carbonic anhydrase. Bicarbonate acts as a key buffer and maintains acid-base balance in biological fluids like blood. For example, during intense exercise, lactic acid can lower pH. Decarboxylation reactions help generate bicarbonate to counteract acidosis, which ensures proper cellular function.

4) Role in Microbial Metabolism

Decarboxylation is also important in microbial metabolism, especially in fermentation and anaerobic respiration. Many microorganisms utilize decarboxylation reactions to generate energy or maintain redox balance under anaerobic conditions. For example:

- In lactic acid bacteria, pyruvate is decarboxylated to produce acetoin and other compounds. These contribute to the flavor of fermented foods like yogurt and cheese.
- Some bacteria decarboxylate amino acids (e.g., lysine to cadaverine) and generate alkaline amines that help neutralize acidic environments and support microbial survival.

These processes have practical implications in food production, biotechnology, and microbial ecology studies.

5) Clinical and Pharmacological Relevance

Disruptions in decarboxylation pathways are linked to several diseases, which makes them targets for therapeutic intervention:

- **Neurological Disorders:** Issues in decarboxylation related to neurotransmitter synthesis are associated with conditions like Parkinson's disease (dopamine deficiency) and depression (serotonin imbalance). Drugs like L-DOPA target these pathways to restore neurotransmitter levels.

- **Metabolic Disorders:** Deficiencies in decarboxylase enzymes, such as pyruvate dehydrogenase (not pyruvate decarboxylase in humans), can impede energy metabolism. This results in conditions like lactic acidosis.

- **Cancer:** Some tumors show altered decarboxylation activity, particularly in polyamine synthesis, which supports rapid cell growth. Inhibitors of decarboxylases, such as ornithine decarboxylase inhibitors, are being explored as potential cancer treatments.

6) Evolutionary Perspective

Decarboxylation reactions are evolutionarily conserved across all life domains, which highlights their fundamental importance. The ability to remove CO_2 from organic molecules allows organisms to optimize metabolic pathways, efficiently generate energy, and produce signaling molecules. The diversity of decarboxylase enzymes and their substrates reflects adaptations to specific ecological niches and physiological needs.

Chapter 11: Enol and Enolate Chemistry

Enol and enolate chemistry represent an important area in organic chemistry. They bridge the reactivity of carbonyl compounds with that of alkenes and carbanions. This duality stems from the ability of carbonyl compounds to exist in equilibrium with their enol tautomers and to form enolate ions under basic conditions. These enol and enolate species are powerful nucleophiles that can participate in a wide range of carbon-carbon bond-forming reactions, which makes them indispensable tools in organic synthesis.

Keto-Enol Tautomerism

Keto-enol tautomerism is a type of constitutional isomerism where two isomers—the keto form and the enol form—are interconverted through the migration of a proton and the relocation of a double bond. The keto form is a carbonyl compound, while the enol form is an alcohol with a double bond adjacent to the alcohol group (an "ene-ol").

Mechanism of Keto-Enol Tautomerism

Tautomerism can be catalyzed by both acids and bases, each of which follows distinct mechanisms that facilitate the conversion between these two tautomeric forms.

Acid-Catalyzed Tautomerism

In acid-catalyzed tautomerism, the process begins with the protonation of the carbonyl oxygen by an acid, which enhances the electrophilicity of the carbonyl carbon. This protonation step is key as it increases the electrophilicity of the carbonyl carbon, facilitating the subsequent deprotonation of the α-carbon to form the enol. Once the carbonyl carbon is activated, the next step involves the deprotonation of the α-carbon, which is the carbon atom adjacent to the carbonyl group. This deprotonation is usually carried out by a base, often a solvent molecule, which abstracts a proton from the α-carbon. The removal of this proton leads to the formation of the enol, characterized by a C=C double bond between the α-carbon and the former carbonyl carbon, and an –OH group replacing the carbonyl oxygen.

This mechanism highlights the role of acid in enhancing the electrophilic character of the carbonyl compound, which facilitates the transition to the enol form. The equilibrium between the keto and enol forms in acid-catalyzed tautomerism is dynamic, influenced by various factors such as the strength of the acid and the solvent used.

Base-Catalyzed Tautomerism

In contrast, base-catalyzed tautomerism initiates with the deprotonation of the α-carbon by a base, which results in the formation of an enolate ion. This enolate ion is a resonance-stabilized species that highlights the nucleophilic character of the α-carbon. The enolate ion can then undergo protonation either at the oxygen atom or the α-carbon, but protonation at the oxygen leads to the enol form, while protonation at the carbon regenerates the keto form. This mechanism underscores the importance of basic conditions in facilitating the conversion to the enol form by first generating a reactive enolate intermediate.

The base-catalyzed pathway is particularly significant in reactions where strong bases are present, as they can effectively remove the proton from the α-carbon. This produces a stable enolate that can readily participate in further reactions. Much like the acid-catalyzed route, the base-catalyzed tautomerism also exists in equilibrium with the keto form, and the position of this equilibrium can be affected by the nature of the base and other reaction conditions.

Equilibrium and Stability

In most cases, the keto form is thermodynamically more stable than the enol form and is therefore the predominant species at equilibrium. This is due to the greater strength of the C=O double bond compared to the C=C double bond. However, in certain cases, the enol form can be stabilized by factors such as conjugation, hydrogen bonding, and aromaticity.

One of the key factors that can stabilize the enol form is conjugation. When the double bond of the enol is conjugated with other π systems, resonance effects can lead to increased stability. This conjugation allows for the delocalization of electrons across the molecule, which lowers the overall energy of the enol form and makes it more favorable under certain conditions.

Another important stabilizing factor is intramolecular hydrogen bonding, particularly prominent in β-dicarbonyl compounds. In these molecules, the enol form can form a stable six-membered ring through hydrogen bonding between the hydroxyl group of the enol and carbonyl oxygen. This intramolecular interaction not only stabilizes the enol form but can also influence the equilibrium position. This makes it more likely to exist in a significant concentration.

In some instances, the formation of the enol can lead to the development of an aromatic system. Aromaticity provides a substantial driving force for enolization because the resulting aromatic compound is more stable than its nonaromatic counterparts. The stability conferred by aromaticity can shift the equilibrium in favor of the enol form, particularly in systems where this transition leads to enhanced stability due to delocalized π electrons.

Significance of Keto-Enol Tautomerism

Halogenation and Electrophilic Addition

Enols are key intermediates in the alpha-halogenation of carbonyl compounds. Under acidic conditions, the enol reacts with halogens (e.g., Br_2) through electrophilic addition to the C=C double bond and places a halogen at the alpha carbon. For example, acetone's enol ($CH_2=C(OH)CH_3$) reacts with Br_2 to form α-bromoacetone (CH_3COCH_2Br). In basic conditions, the enolate drives reactions like the haloform reaction, where exhaustive halogenation of methyl ketones (e.g., acetone) leads to cleavage and produces a carboxylate and a haloform (e.g., $CHBr_3$). This reaction is historically significant for identifying methyl ketones and has synthetic utility in preparing carboxylic acids.

Rearrangements and Cyclizations

Keto-enol tautomerism facilitates rearrangement reactions by allowing interconversion between structural isomers. For example, in the Favorskii rearrangement, an α-halo ketone, under basic conditions, forms an enolate that rearranges to a carboxylic acid or ester through a cyclopropanone intermediate, and tautomerism enables the necessary bond shifts. Similarly, tautomerism is important in cyclization reactions, such as the formation of heterocycles (e.g., pyrazoles, isoxazoles) from 1,3-dicarbonyls, where the enol form participates in nucleophilic attack or condensation steps.

Stabilization of Reaction Intermediates

The enol or enolate often serves as a stabilized intermediate in reactions, lowering transition state energies. In enzymatic reactions or organocatalytic processes, the enol's ability to form hydrogen bonds or coordinate with catalysts enhances reaction efficiency. For instance, in asymmetric aldol reactions, chiral catalysts stabilize the enolate in a specific conformation and enable stereoselective C-C bond formation.

Enzymatic Catalysis

Enzymes often exploit keto-enol tautomerism to facilitate metabolic transformations. For example, in glycolysis, the enzyme enolase catalyzes the dehydration of 2-phosphoglycerate to phosphoenolpyruvate (PEP), an enol that is highly reactive due to its phosphate group and C=C bond. The enol form of PEP is key for its role as a high-energy phosphate donor in ATP synthesis. Similarly, isomerases such as triose phosphate isomerase catalyze the interconversion of dihydroxyacetone phosphate (keto) and glyceraldehyde 3-phosphate (enol-like intermediate), a key step in carbohydrate metabolism. The enzyme stabilizes the enediol-like transition state through active site residues and lowers the activation energy.

DNA and RNA Stability

Keto-enol tautomerism in nucleic acid bases, such as guanine and thymine, affects their hydrogen-bonding properties and base-pairing fidelity. Under rare conditions, bases can tautomerize to enol or iminol forms, which results in tautomeric shifts that cause mispairing during DNA replication. For example, the enol form of thymine may pair with guanine instead of adenine and introduce mutations. These rare tautomeric forms are implicated in spontaneous mutations and contribute to genetic diversity and, in some cases, disease.

Metabolite Reactivity

Many metabolites, such as pyruvic acid ($CH_3COCOOH$), exist in keto-enol equilibrium. This influences their biochemical roles. The enol form of pyruvate ($CH_2=C(OH)COOH$) is more nucleophilic and can participate in reactions like decarboxylation, but transamination typically involves the keto form of α-keto acids. Similarly, keto acids in the citric acid cycle rely on tautomerism to modulate reactivity, which enables transformations catalyzed by enzymes like isocitrate dehydrogenase.

Drug Design and Pharmacology

Keto-enol tautomerism affects the activity and metabolism of pharmaceutical compounds. For example, barbiturates, used as sedatives, exist in keto and enol forms, and the enol form contributes to their acidity and solubility. This influences their pharmacokinetic properties. The tautomeric state can also affect a drug's binding to its target, as enzymes or receptors may preferentially recognize one tautomer. In drug design, knowledge of tautomerism is important to optimize bioavailability, stability, and specificity.

Enolate Formation and Reactivity

Enolates are negatively charged species formed by the deprotonation of an α-carbon of a carbonyl compound. They are powerful nucleophiles and versatile intermediates in organic synthesis.

Formation of Enolates

Enolates are formed when a carbonyl compound is treated with a strong base. The choice of base is important for controlling the regioselectivity and stereoselectivity of enolate formation.

The formation of enolates begins with the deprotonation of a carbonyl compound, which can occur through various strong bases. Commonly employed bases include sodium hydride (NaH), lithium diisopropylamide (LDA), and sodium hexamethyldisilazide (NaHMDS), which are strong, non-nucleophilic bases suitable for generating enolates selectively. These strong, nonselective bases can deprotonate most carbonyl compounds. However, their nonselective nature can lead to the formation of a mixture of enolates, particularly when the carbonyl compound possesses multiple α-carbons. This lack of selectivity complicates the synthesis of specific enolates, as each deprotonation site can yield different products.

Bulky, Non-Nucleophilic Bases

To achieve greater control over enolate formation, chemists often utilize bulky, non-nucleophilic bases such as lithium diisopropylamide (LDA) and potassium tert-butoxide (KOtBu). These bases are advantageous because their size and steric hindrance allow them to selectively deprotonate the more accessible, less-substituted α-carbon, leading to kinetic enolate formation. For instance, LDA is particularly effective for generating kinetic enolates due to its ability to rapidly deprotonate less substituted carbons while minimizing side reactions. This selective deprotonation is vital for applications that rely on the formation of specific enolates for subsequent reactions.

Kinetic vs. Thermodynamic Enolates

The presence of multiple α-carbons in a carbonyl compound can lead to the formation of different enolates, categorized as kinetic and thermodynamic enolates. The distinction between these two types highlights the importance of reaction conditions in determining the outcome of enolate formation.

Kinetic Enolate

The kinetic enolate is formed under conditions that favor rapid, irreversible deprotonation. Usually, this involves the use of a strong, bulky base like LDA at low temperatures. Under these conditions, the reaction proceeds quickly and favors the formation of the less substituted enolate. This is because the kinetic enolate forms more quickly due to lower activation energy, typically at the less-substituted, more accessible α-carbon, despite being less stable overall. The kinetic enolate is often the preferred product when immediate reactivity is desired, such as in certain synthetic routes.

Thermodynamic Enolate

In contrast, the thermodynamic enolate is formed under conditions that allow for equilibrium to be established. This process usually employs a weaker base, such as sodium ethoxide (NaOEt), at higher temperatures. Under these conditions, the reaction can proceed to completion and allows for the more stable, substituted enolate to form. The thermodynamic enolate is usually favored when the goal is to produce a more stable product, as the stability of the enolate is enhanced by greater substitution at the α-carbon.

Structure and Stability of Enolates

The negative charge in an enolate is delocalized between the α-carbon and the oxygen atom. This delocalization contributes to the stability of the enolate and makes both the α-carbon and the oxygen atom nucleophilic.

Reactivity of Enolates

Enolates are ambident nucleophiles, which means that they can react with electrophiles at two different sites: the α-carbon and the oxygen atom.

- C-Alkylation: Reaction at the α-carbon leads to the formation of a new carbon-carbon bond. This is the most common and synthetically useful mode of reactivity.
- O-Alkylation: Reaction at the oxygen atom leads to the formation of an enol ether. This is generally less desirable than C-alkylation.

The regioselectivity of enolate alkylation can be influenced by factors such as the nature of the electrophile, the solvent, and the counterion.

Aldol Reactions and Claisen Condensations

Aldol reactions and Claisen condensations are two of the most important carbon-carbon bond-forming reactions in organic chemistry. They involve the reaction of an enolate with a carbonyl compound.

Aldol Reaction

The aldol reaction is the reaction of an enolate with an aldehyde or ketone to form a β-hydroxyaldehyde or β-hydroxyketone (an "aldol"). This reaction is usually catalyzed by a base or an acid.

Base-Catalyzed Aldol Reaction

- Enolate Formation: A base removes a proton from the α-carbon of one carbonyl compound and forms an enolate.
- Nucleophilic Attack: The enolate attacks the carbonyl carbon of another carbonyl compound and forms a tetrahedral intermediate.
- Protonation: The alkoxide intermediate is protonated to give the β-hydroxyaldehyde or β-hydroxyketone.

Acid-Catalyzed Aldol Reaction

- Enol Formation: An acid catalyzes the formation of an enol from one carbonyl compound.
- Electrophilic Attack: The enol acts as a nucleophile and attacks the protonated carbonyl group of another carbonyl compound.
- Deprotonation: A proton is removed to give the β-hydroxyaldehyde or β-hydroxyketone.

The aldol product can undergo dehydration (loss of water) to form an α,β-unsaturated aldehyde or ketone. This dehydration is usually promoted by acid or base and leads to the formation of a conjugated system, which is thermodynamically more stable.

Crossed Aldol Reaction

If two different carbonyl compounds are used in an aldol reaction, a mixture of products can be formed. This is known as a crossed aldol reaction. To minimize the formation of unwanted products, one of the carbonyl compounds is often used in excess, or a carbonyl compound that cannot form an enolate is used.

Intramolecular Aldol Reaction

If a molecule contains two carbonyl groups and at least one α-hydrogen, an intramolecular aldol reaction can occur. This reaction is particularly useful when you want to form cyclic compounds.

Claisen Condensation

The Claisen condensation is the reaction of two esters to form a β-keto ester. This reaction is usually carried out using a strong base, such as sodium ethoxide (NaOEt).

The mechanism of the Claisen condensation is similar to that of the aldol reaction:

- Enolate Formation: A base removes a proton from the α-carbon of one ester molecule and forms an enolate.
- Nucleophilic Attack: The enolate attacks the carbonyl carbon of another ester molecule and forms a tetrahedral intermediate.
- Elimination: The ethoxide group is eliminated, regenerates the carbonyl group, and forms the β-keto ester.
- Deprotonation: The β-keto ester is deprotonated by the base and forms a stable enolate anion. This deprotonation is necessary to drive the equilibrium toward product formation.

The Claisen condensation is reversible, and the equilibrium usually favors the starting materials. However, if the β-keto ester product has an acidic α-proton, it can be deprotonated by the base, which forms a stable enolate anion. This deprotonation shifts the equilibrium toward product formation.

Dieckmann Condensation

The Dieckmann condensation is an intramolecular Claisen condensation. This reaction is particularly useful when you want to form cyclic β-keto esters. In the Dieckmann condensation, a diester undergoes deprotonation at one of its α-carbons by a strong base, generating an enolate ion. This enolate then attacks the carbonyl carbon of another ester group within the same molecule, which results in the formation of a cyclic intermediate. Upon acidic or basic work-up, this intermediate yields a cyclic β-keto ester, which is characterized by the presence of both a ketone and an ester functional group within a ring system.

The ability to form cyclic β-keto esters through the Dieckmann condensation is particularly advantageous in synthetic chemistry. These compounds can serve as precursors for various biologically active molecules and are also useful in the synthesis of complex natural products. Additionally, the Dieckmann condensation exemplifies the strategic use of intramolecular reactions to enhance reaction efficiency and selectivity.

Conjugate (Michael) Additions

Conjugate addition, also known as the Michael addition, is the addition of a nucleophile to the β-carbon of an α,β-unsaturated carbonyl compound. This reaction is a powerful method used to form carbon-carbon bonds and introduce new functional groups into a molecule.

Mechanism of Conjugate Addition

The mechanism of conjugate addition involves the following steps:

1) Nucleophilic Attack: The nucleophile attacks the β-carbon of the α,β-unsaturated carbonyl compound, which forms an enolate intermediate.

2) Protonation: The enolate intermediate is protonated to give the conjugate addition product.

The α,β-unsaturated carbonyl compound acts as an electrophile, with the β-carbon being the electrophilic site. The nucleophile can be a variety of species, including:

- Enolates: Enolates are commonly used as nucleophiles in Michael additions. This reaction is known as the Michael reaction.
- Amines: Amines can also act as nucleophiles in Michael additions.
- Thiols: Thiols are good nucleophiles for Michael additions, particularly in biological systems.
- Cyanide: Cyanide ions can add to α,β-unsaturated carbonyl compounds.

Michael Reaction

The Michael reaction is the conjugate addition of an enolate to an α,β-unsaturated carbonyl compound. This reaction is a powerful method used to form carbon-carbon bonds and is widely used in organic synthesis.

The Michael reaction is usually carried out using a base, such as sodium ethoxide (NaOEt) or potassium tert-butoxide (KOtBu), which acts as a reagent rather than a true catalyst. The base promotes the formation of the enolate from the donor carbonyl compound.

Factors that Affect Conjugate Addition

Several factors can influence the rate and regioselectivity of conjugate addition reactions:

- Nature of the Nucleophile: The reactivity of the nucleophile is an important factor. Stronger nucleophiles tend to react faster.
- Nature of the α,β-Unsaturated Carbonyl Compound: The electrophilicity of the α,β-unsaturated carbonyl compound is also important. EWGs on the carbonyl compound increase its electrophilicity and make it more reactive.
- Steric Effects: Steric hindrance can affect the regioselectivity of the reaction. Bulky nucleophiles may prefer to attack the less hindered β-carbon.
- Solvent: The solvent can also influence the rate and regioselectivity of the reaction.

Robinson Annulation

The Robinson annulation is a powerful method used to form six-membered rings. The first phase of the Robinson annulation involves a Michael addition, where a nucleophilic attack occurs. This process begins with the formation of an enolate ion, generated when a strong base (such as sodium hydride or potassium tert-butoxide) deprotonates a carbonyl compound at its α-position. The enolate ion that results is a resonance-stabilized nucleophile that is key for the subsequent nucleophilic attack.

In the Michael addition, the enolate ion then adds to the β-carbon of an α,β-unsaturated carbonyl compound, which acts as the electrophile. This step is important as it establishes a new carbon-carbon bond, which effectively extends the carbon chain and produces a diketone intermediate. This diketone serves as the foundation for the next step of the reaction and sets up the molecular framework necessary for intramolecular aldol condensation.

The second phase of the Robinson annulation transforms the diketone intermediate into a six-membered ring through an intramolecular aldol addition followed by a dehydration (condensation) step. This step begins with the deprotonation of one of the carbonyl groups in the diketone by a base, which generates another enolate ion.

This newly formed enolate then attacks the second carbonyl group within the same molecule. The result is the formation of a β-hydroxycarbonyl compound, which features a six-membered ring with a hydroxyl group adjacent to a carbonyl group. The final transformation occurs when this β-hydroxycarbonyl intermediate loses a molecule of water through dehydration and yields an α,β-unsaturated carbonyl system. It usually results in a cyclohexenone ring.

The intramolecular nature of this aldol condensation favors ring formation and minimizes the potential for competing side reactions. This inherent selectivity ensures that the desired six-membered ring product is obtained efficiently.

The Robinson annulation is renowned for several compelling reasons:

- **Versatility**: This reaction efficiently constructs six-membered rings, a prevalent motif in organic molecules. Its ability to generate complex cyclic structures is invaluable in synthetic organic chemistry.
- **Stereoselectivity**: The reaction often produces specific stereoisomers, which is important in the synthesis of biologically active compounds. The controlled formation of stereocenters enhances the utility of the Robinson annulation in pharmaceuticals.
- **Synthetic Utility**: The Robinson annulation serves as a key tool for building intricate structures, particularly those found in natural products. Its capability to create multiple bonds and functional groups in a single reaction sequence streamlines synthetic pathways.

Chapter 12: Oxidation and Reduction

Oxidation and reduction, often referred to as redox reactions, are fundamental processes in organic chemistry. They involve the transfer of electrons between reactants, which results in changes in oxidation states. Oxidation is defined as the loss of electrons or an increase in oxidation state, while reduction is defined as the gain of electrons or a decrease in oxidation state. These reactions interconvert functional groups and construct complex molecules.

Oxidizing Agents

Oxidizing agents are substances that accept electrons, which causes another species to be oxidized. In organic chemistry, oxidation usually increases the number of bonds to oxygen or decreases the number of bonds to hydrogen. Several oxidizing agents are commonly used to transform alcohols, aldehydes, and other functional groups.

Pyridinium Chlorochromate (PCC)

Pyridinium chlorochromate (PCC) is a mild oxidizing agent that selectively oxidizes primary alcohols to aldehydes and secondary alcohols to ketones. It is a complex of chromium trioxide (CrO_3), pyridine (C_5H_5N), and hydrochloric acid (HCl).

- Structure: $[C_5H_5NH^+][CrO_3Cl^-]$
- Preparation: Chromium trioxide is carefully added to a solution of pyridine in hydrochloric acid.
- Solvent: Usually used in dichloromethane (CH_2Cl_2) as a solvent.

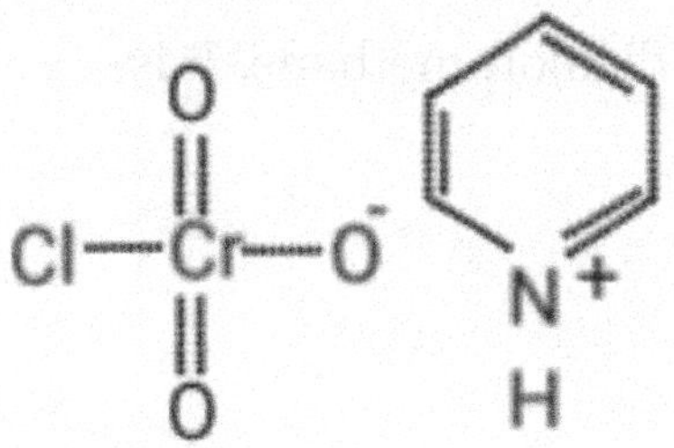

Fig 36: Structure of pyridinium chlorochromate.

Mechanism of Oxidation with PCC

1) Formation of Chromate Ester: The alcohol reacts with PCC to form a chromate ester. This involves the coordination of the alcohol oxygen to the chromium atom.

2) Elimination: A proton is abstracted from the carbon atom that bears the oxygen, which results in the elimination of a chromium(IV) intermediate and the formation of the carbonyl compound (aldehyde or ketone).

Advantages of PCC

- Mild and Selective: PCC is a mild oxidizing agent that does not usually over-oxidize aldehydes to carboxylic acids. This is a significant advantage over stronger oxidizing agents like CrO_3 or $KMnO_4$.
- Good Yields: PCC generally provides good yields of aldehydes and ketones.
- Relatively Easy to Use: PCC is relatively easy to handle and use in the laboratory.

Limitations of PCC

- Stoichiometric Reagent: PCC is a stoichiometric reagent, which means that one mole of PCC is required for each mole of alcohol oxidized.
- Toxic: Chromium compounds are toxic and must be handled with care.

Chromium Trioxide (CrO_3)

Chromium trioxide (CrO_3), also known as chromic acid, is a powerful oxidizing agent that can oxidize primary alcohols to carboxylic acids and secondary alcohols to ketones. It is usually used in aqueous sulfuric acid (H_2SO_4) or acetic acid (CH_3COOH) as a solvent.

- Jones Reagent: A solution of CrO_3 in aqueous sulfuric acid is known as the Jones reagent. It is a very effective oxidizing agent used to convert primary alcohols to carboxylic acids.
- Collins Reagent: A complex of CrO_3 with pyridine in dichloromethane. It is similar to PCC but more difficult to prepare.

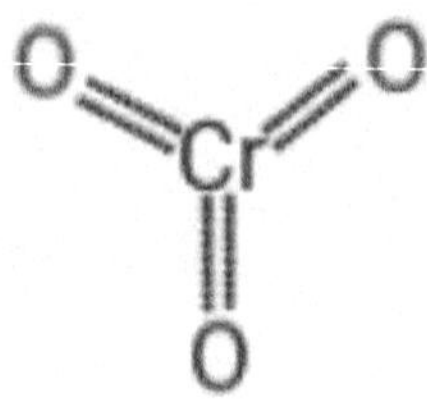

Fig 37: Structure of chromium trioxide.

Mechanism of Oxidation with CrO_3

The mechanism of oxidation with CrO_3 is similar to that of PCC, which involves the formation of a chromate ester intermediate followed by elimination. However, because CrO_3 is a stronger oxidizing agent, the reaction usually proceeds further than with PCC.

1) Formation of Chromate Ester: The alcohol reacts with CrO_3 to form a chromate ester.

2) Elimination: A proton is abstracted from the carbon atom that bears the oxygen, which results in the formation of a carbonyl compound and reduction of chromium to a chromium(IV) or chromium(III) intermediate.

3) Further Oxidation (for Primary Alcohols): If a primary alcohol is used, the aldehyde intermediate can be further oxidized to a carboxylic acid. This involves the addition of water to the aldehyde, followed by oxidation of the resulting gem-diol.

Limitations of CrO_3

- Strong Oxidizing Agent: CrO_3 is a strong oxidizing agent that can over-oxidize aldehydes to carboxylic acids.
- Harsh Conditions: The reaction conditions are often harsh, which can lead to unwanted side reactions.
- Toxic: Chromium compounds are toxic and must be handled with care.
- Not Suitable for Acid-Sensitive Compounds: The strongly acidic conditions can be problematic for molecules that contain acid-sensitive functional groups.

Potassium Permanganate ($KMnO_4$)

Potassium permanganate ($KMnO_4$) is a strong oxidizing agent that can oxidize a wide variety of organic compounds. It is usually used in aqueous solution, either under acidic or basic conditions.

- Acidic Conditions: Under acidic conditions, $KMnO_4$ oxidizes primary alcohols to carboxylic acids and secondary alcohols to ketones.
- Basic Conditions: Under basic conditions, $KMnO_4$ can oxidize alkenes to syn-diols (dihydroxylation).

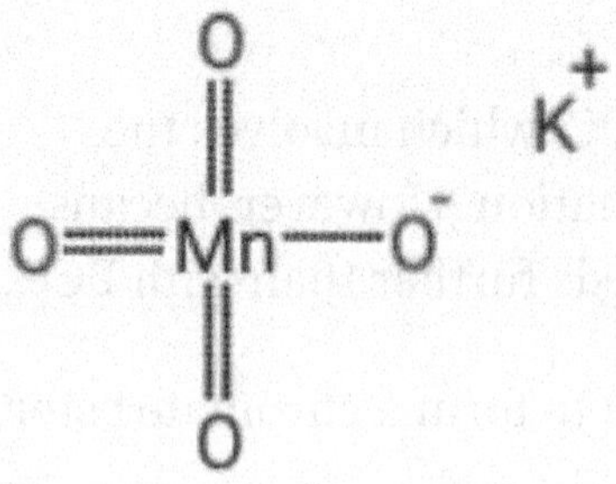

Fig 38: Structure of potassium permanganate.

Mechanism of Oxidation with $KMnO_4$

The mechanism of oxidation with $KMnO_4$ is complex and depends on the reaction conditions. In general, it involves the formation of a manganate ester intermediate, followed by oxidative cleavage and formation of the oxidized product along with manganese dioxide (MnO_2) or other manganese species, depending on the reaction conditions.

Limitations of $KMnO_4$

- Strong Oxidizing Agent: $KMnO_4$ is a strong oxidizing agent that can lead to over-oxidation and unwanted side reactions.
- Harsh Conditions: The reaction conditions can be harsh, which can limit the oxidizing agent's use with sensitive compounds.
- Formation of MnO_2: The formation of manganese dioxide (MnO_2) as a byproduct can make product isolation difficult.
- Not Suitable for Acid-Sensitive Compounds: Acidic conditions are often employed, which can be problematic.

Reducing Agents

Reducing agents are substances that donate electrons, which causes another species to be reduced. In organic chemistry, reduction usually increases the number of bonds to hydrogen or decreases the number of bonds to oxygen. Several reducing agents are commonly used to reduce carbonyl compounds, carboxylic acids, and other functional groups.

Lithium Aluminum Hydride ($LiAlH_4$)

Lithium aluminum hydride ($LiAlH_4$) is a powerful reducing agent that can reduce a wide variety of functional groups, including aldehydes, ketones, carboxylic acids, esters, amides, and nitriles. It is a source of hydride ions (H^-), which act as nucleophiles in reduction reactions.

- Structure: $LiAlH_4$
- Solvent: Usually used in anhydrous ether solvents such as diethyl ether (Et_2O) or tetrahydrofuran (THF).
- Highly Reactive: $LiAlH_4$ is a very reactive reagent and must be handled with care. It reacts violently with water and other protic solvents.

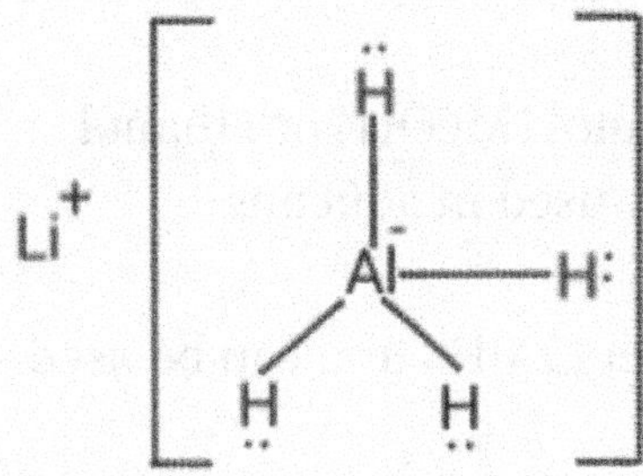

Fig 39: Structure of lithium aluminum hydride.

Mechanism of Reduction with $LiAlH_4$

1) Nucleophilic Attack: The hydride ion (H^-) from $LiAlH_4$ attacks the electrophilic carbon atom of the carbonyl group, which forms an alkoxide intermediate.

2) Coordination: The aluminum species coordinates with the oxygen atom of the carbonyl during the initial complex formation, and later with the resulting alkoxide intermediate.

3) Further Reduction (for Carboxylic Acids and Esters): For carboxylic acids and esters, the reduction process involves multiple steps. The initial reduction forms an aldehyde intermediate, which is then further reduced to an alcohol.

4) Protonation: After the reduction is complete, the reaction mixture is treated with dilute acid to protonate the alkoxide intermediate and give the alcohol product.

Limitations of $LiAlH_4$

- Strong Reducing Agent: $LiAlH_4$ can reduce a wide variety of functional groups, which can be a limitation if selectivity is desired.
- Highly Reactive: $LiAlH_4$ is very reactive and must be handled with care.

- Reacts with Protic Solvents: $LiAlH_4$ reacts violently with water and other protic solvents.
- Expensive: $LiAlH_4$ is relatively expensive.

Sodium Borohydride ($NaBH_4$)

Sodium borohydride ($NaBH_4$) is a milder reducing agent than $LiAlH_4$. It can reduce aldehydes and ketones to alcohols, but it does not usually reduce carboxylic acids, esters, or amides.

- Structure: $NaBH_4$
- Solvent: Usually used in protic solvents such as methanol (MeOH) or ethanol (EtOH); it reacts slowly in water and is less commonly used in aqueous conditions.
- Less Reactive than $LiAlH_4$: $NaBH_4$ is less reactive than $LiAlH_4$ and can be used in protic solvents.

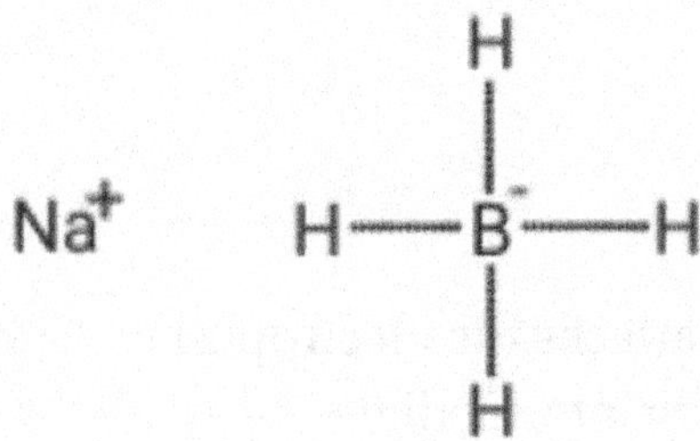

Fig 40: Structure of sodium borohydride.

Mechanism of Reduction with $NaBH_4$

The mechanism of reduction with $NaBH_4$ is similar to that of $LiAlH_4$, which involves the nucleophilic attack of a hydride ion on the carbonyl carbon.

1) Nucleophilic Attack: The hydride ion (H^-) from $NaBH_4$ attacks the electrophilic carbon atom of the carbonyl group and forms an alkoxide intermediate.

2) Coordination: The oxygen atom of the alkoxide intermediate remains associated with the boron center as part of the borate complex before protonation.

3) Protonation: After the reduction is complete, the reaction mixture is treated with dilute acid to protonate the alkoxide intermediate and give the alcohol product.

Advantages of $NaBH_4$

- Milder Reducing Agent: $NaBH_4$ is a milder reducing agent than $LiAlH_4$ and is more selective.
- Can be Used in Protic Solvents: $NaBH_4$ can be used in alcohol solvents; it reacts slowly in water and is less commonly used in fully aqueous systems.
- Less Expensive: $NaBH_4$ is less expensive than $LiAlH_4$.

Limitations of $NaBH_4$

- Does Not Reduce Carboxylic Acids, Esters, or Amides: $NaBH_4$ does not usually reduce carboxylic acids, esters, or amides.
- Slower Reaction Rate: The reaction rate is slower than with $LiAlH_4$.

Catalytic Hydrogenation

Catalytic hydrogenation is the addition of hydrogen (H_2) to a molecule in the presence of a metal catalyst. This reaction is commonly used to reduce alkenes, alkynes, and carbonyl compounds.

- Catalysts: Common catalysts include palladium (Pd), platinum (Pt), nickel (Ni), and rhodium (Rh). The catalyst is usually supported on a high-surface-area material such as carbon (C).
- Solvent: Usually used in alcohol solvents such as ethanol (EtOH), methanol (MeOH), or ethyl acetate (EtOAc).

Mechanism of Catalytic Hydrogenation

1) Adsorption: The hydrogen molecule and the organic molecule adsorb onto the surface of the metal catalyst.

2) Activation: The hydrogen molecule is activated by the metal catalyst, which breaks the H-H bond and forms metal-hydrogen bonds.

3) Transfer: The hydrogen atoms are transferred to the organic molecule, which reduces the double or triple bond.

4) Desorption: The reduced product desorbs from the surface of the metal catalyst.

Selectivity in Catalytic Hydrogenation

Catalytic hydrogenation can be selective for certain functional groups. For example, alkenes can be selectively reduced in the presence of carbonyl groups through the use of a poisoned catalyst such as Lindlar's catalyst (palladium on calcium carbonate poisoned with lead acetate).

Limitations of Catalytic Hydrogenation:

- Requires High Pressure: Catalytic hydrogenation may require elevated hydrogen pressure, but many reductions proceed efficiently at atmospheric pressure depending on the substrate and catalyst used.
- Can Reduce Multiple Functional Groups: Catalytic hydrogenation can reduce multiple functional groups, which can be a limitation if selectivity is desired.
- Stereochemistry: The stereochemistry of the product can be difficult to control.

Selectivity in Redox Reactions

Selectivity is an important consideration in organic synthesis. When multiple functional groups are present in a molecule, it is often necessary to selectively oxidize or reduce one functional group and leave others untouched.

Chemoselectivity

Chemoselectivity refers to the ability of a reagent to react preferentially with one functional group over another.

- PCC vs. CrO_3: PCC is chemoselective for oxidizing alcohols to aldehydes or ketones without oxidizing other functional groups, while CrO_3 is less selective.
- $NaBH_4$ vs. $LiAlH_4$: $NaBH_4$ is chemoselective for reducing aldehydes and ketones without reducing carboxylic acids, esters, or amides, while $LiAlH_4$ is less selective.
- Catalytic Hydrogenation: When catalyst and reaction conditions are carefully chosen, catalytic hydrogenation can be made chemoselective for reducing certain functional groups.

Regioselectivity

Regioselectivity refers to the ability of a reagent to react preferentially at one site over another within the same molecule.

- Oxidation of Unsymmetrical Alcohols: The oxidation of a molecule with multiple alcohol groups can be regioselective if one hydroxyl group is more hindered or more reactive than the other.
- Reduction of Unsymmetrical Ketones: The reduction of an unsymmetrical ketone can be regioselective if one of the carbonyl groups is more hindered than the other.

Stereoselectivity

Stereoselectivity refers to the ability of a reagent to form one stereoisomer preferentially over another.

- Catalytic Hydrogenation: Catalytic hydrogenation can be stereoselective if the catalyst is chiral or the reaction is carried out in the presence of a chiral auxiliary.
- Reduction of Cyclic Ketones: The reduction of a cyclic ketone can be stereoselective. The hydride ion attacks preferentially from one face of the ring.

Application to Synthesis

Oxidation and reduction reactions are essential tools in organic synthesis. They are used to interconvert functional groups, construct complex molecules, and introduce chirality.

Functional Group Interconversion

Oxidation and reduction reactions can be used to interconvert a variety of functional groups.

- Alcohols to Aldehydes/Ketones: PCC, CrO_3
- Aldehydes/Ketones to Alcohols: $NaBH_4$, $LiAlH_4$, Catalytic Hydrogenation
- Alkenes to Alkanes: Catalytic Hydrogenation
- Alcohols to Carboxylic Acids: $KMnO_4$, CrO_3

Construction of Complex Molecules

Oxidation and reduction reactions are often used in multistep syntheses to construct complex molecules.

- Corey-House Synthesis: The Corey-House synthesis uses lithium diorganocuprate reagents to couple with alkyl halides to form higher alkanes via a carbon–carbon bond-forming reaction.

- Grignard Reaction: The Grignard reaction uses Grignard reagents to add alkyl groups to carbonyl compounds.

Introduction of Chirality

Oxidation and reduction reactions can be used to introduce chirality into a molecule.

- Sharpless Epoxidation: The Sharpless epoxidation is a stereoselective method used to epoxidize allylic alcohols via a titanium catalyst and a chiral tartrate ester.
- CBS Reduction: The Corey-Bakshi-Shibata (CBS) reduction is a stereoselective method used to reduce ketones to alcohols with a chiral oxazaborolidine catalyst.

Chapter 13: Synthesis and Multistep Synthesis

Organic synthesis involves the construction of organic molecules from simpler building blocks. It is a central discipline in chemistry, with applications that range from the development of new pharmaceuticals and materials to the understanding of complex biological processes. Multistep synthesis involves the sequential execution of multiple chemical reactions to achieve the desired target molecule. The design of efficient and effective multistep syntheses requires a deep knowledge of organic reactions, functional group chemistry, and strategic planning.

Strategy for Retrosynthetic Analysis

Retrosynthetic analysis is a problem-solving technique used to plan organic syntheses. It requires a chemist to work backward from the target molecule to identify simpler starting materials and reaction sequences that can be used to construct the target. The key idea is to break down the target molecule into smaller, more manageable fragments by mentally "disconnecting" bonds.

Key Concepts in Retrosynthetic Analysis

- Target Molecule: The molecule that you want to synthesize.
- Starting Materials: Commercially available or readily accessible compounds that will be used as the starting points for the synthesis.
- Synthetic Equivalent: A reagent or reaction that can be used to achieve a specific transformation.
- Disconnection: The mental breaking of a bond in the target molecule.
- Synthon: An idealized fragment of a molecule that results from a disconnection. Synthons may be charged or neutral and are theoretical constructs used to guide the selection of reagents in synthesis.
- Reagent: A real chemical substance that can be used to generate or react with a synthon.
- Transform: A chemical reaction that converts one functional group into another.
- Retron: A minimal structural element that can be associated with a specific transform.
- Arrow: A retrosynthetic arrow (⇒) indicates a retrosynthetic step, i.e., a step that is performed mentally in the reverse direction of the actual synthesis.

Steps in Retrosynthetic Analysis

Step 1: Analyze the Target Molecule

The first step in retrosynthetic analysis is to carefully analyze the target molecule. Next, identify its functional groups, stereocenters, and any unusual structural features. It's important to understand the molecular architecture, as these elements dictate the reactivity and the types of reactions that can be employed. Once they recognize the overall structure, chemists can establish a foundation for the subsequent steps in the analysis.

Step 2: Identify Key Bonds to Disconnect

Next, the focus shifts to key bonds within the target molecule that must be identified so they can be disconnected. The goal is to simplify the molecule and facilitate the pathway toward readily available starting materials. Several factors should be considered during this process:

- **Functional Group Chemistry**: Bonds adjacent to functional groups must often be strategically disconnected, as these locations are usually sites of reactivity. This can lead to simpler intermediates that can be further elaborated.
- **Symmetry**: It is helpful to identify bonds that create symmetrical intermediates, as symmetrical compounds can simplify the synthesis and minimize the number of different reagents needed.
- **Ring Systems**: If the target molecule contains ring structures, disconnections are often made at bonds that can lead to precursors capable of undergoing ring-closing reactions, facilitating the synthesis of cyclic compounds. Such transformations often provide a more straightforward synthetic route.
- **Stereochemistry**: The stereochemistry of the target molecule should also be taken into account. It's important to choose disconnections that allow for the controlled introduction of stereocenters. This will facilitate the synthesis of the desired stereoisomer.

Step 3: Propose Synthons

Once key bonds have been identified for disconnection, the next step is to propose the corresponding synthons. Synthons are idealized fragments that represent the theoretical building blocks needed for synthesis. Even though they may not always be stable or readily available, they serve as a conceptual guide for the synthetic pathway. Once they identify these synthons, chemists can outline potential routes to achieve the target molecule.

Step 4: Identify Reagents and Reactions

With the synthons in hand, the next step is to identify the appropriate reagents and reactions that can generate or react with these fragments. The chosen reagents and reactions should be practical, efficient, and selective. This ensures that the synthesis can be completed with minimal side reactions. This step is vital to determine the feasibility of the proposed synthetic route.

Step 5: Repeat the Process

The disconnection process should be repeated iteratively until the analysis leads to readily available starting materials. Each cycle of disconnection and identification of synthons and reagents should progressively simplify the target molecule, which will guide chemists to the most practical synthetic pathway.

Step 6: Write the Forward Synthesis

Finally, once a comprehensive retrosynthetic analysis has been completed, the last step is to write out the forward synthesis. The synthetic route must be detailed, including the reagents and conditions for each step. The forward synthesis serves as a road map for the experimental work required to construct the target molecule from its starting materials.

Guidelines for Retrosynthetic Analysis

Start with the Most Complex Part of the Molecule

A fundamental guideline in retrosynthetic analysis is to begin with the most complex part of the target molecule. When they focus on the most challenging structural features first, chemists can simplify the synthesis process. This approach allows for the identification of key bonds that, when disconnected, can lead to strategic intermediates from which the complex portion of the molecule can be efficiently constructed. If complexity is tackled early in the analysis, this often facilitates the overall synthetic strategy and reduces the likelihood of unforeseen difficulties later on.

Use Functional Group Interconversions (FGIs)

FGIs are essential tools in retrosynthetic analysis. These conversions allow chemists to transform functional groups into more useful or reactive forms, which enhances the molecule's synthetic accessibility. When they strategically plan FGIs, chemists can create intermediates that are better suited for subsequent transformations. This flexibility is particularly valuable in navigating the complexities of organic synthesis, where the reactivity of functional groups can dictate the success of a reaction.

Consider Protecting Groups

In many synthetic pathways, certain functional groups may react in ways that interfere with desired transformations. To mitigate this issue, the use of protecting groups becomes vital. Protecting groups can temporarily mask reactive functional groups, which allows chemists to carry out specific reactions without unwanted side reactions occurring. When they incorporate protecting groups into retrosynthetic analysis, chemists can maintain control over the reaction conditions and ensure that the desired transformations proceed smoothly.

Look for Key Reactions

An important aspect of retrosynthetic analysis is the identification of key reactions that can generate multiple bonds or stereocenters in a single step. These reactions can significantly streamline the synthesis process and allow for the rapid construction of complex molecular architectures. When they leverage powerful reactions, such as the Robinson annulation or Michael addition, chemists can efficiently build the desired structure and minimize the number of synthetic steps required.

Keep It Simple

Simplicity is a core principle in retrosynthetic analysis. Chemists should aim for the shortest and most efficient synthesis possible and avoid unnecessary steps or reagents. A straightforward synthetic route not only reduces the overall time and effort required for synthesis but also minimizes the potential for errors and side reactions. When they prioritize simplicity, chemists can enhance the feasibility of their synthetic plans and improve the likelihood of successful outcomes.

Be Realistic

It is essential to approach retrosynthetic analysis with a realistic mindset. Chemists must consider the practicality of each step in the proposed synthesis, including the cost of reagents, the availability of starting materials, and the potential for side reactions. Through a careful evaluation of these factors, chemists can develop a more accurate and achievable synthetic plan. A realistic approach ensures that the proposed synthesis is not only theoretically sound but also practically feasible, which results in successful experimental outcomes.

Examples of Retrosynthetic Analysis

Example 1: Synthesis of 2-Cyclohexenone

- Target Molecule: 2-Cyclohexenone.
- Disconnection: Disconnect the bond between the α-carbon of the enolate and the β-carbon of the α,β-unsaturated ketone, corresponding to a Michael addition followed by intramolecular aldol condensation (Robinson annulation).
- Synthons: Enolate and α,β-unsaturated ketone.
- Reagents: Methyl vinyl ketone (MVK) and a ketone enolate.
- Starting Materials: Cyclohexanone and MVK.

Example 2: Synthesis of a Grignard Reagent

- Target Molecule: Benzyl alcohol.
- Disconnection: Disconnect the bond between the benzyl carbon and the hydroxyl-bearing carbon through a Grignard reaction.
- Synthons: Benzaldehyde and a Grignard reagent (phenylmagnesium bromide).
- Reagents: Magnesium, benzyl bromide, and benzaldehyde.
- Starting Materials: Benzyl bromide and benzaldehyde.

FGIs are essential tools in organic synthesis and allow chemists to manipulate the reactivity and properties of molecules.

Common Functional Group Interconversions

Oxidation:

- Alcohol to Aldehyde/Ketone: PCC, Swern oxidation, Dess-Martin periodinane.
- Alcohol to Carboxylic Acid: $KMnO_4$, CrO_3 in H_2SO_4 (Jones reagent).
- Alkene to Epoxide: mCPBA, peroxyacetic acid.

$$H_3C-CH_2OH \longrightarrow H_3C-CH{=}O$$

Ethanol Acetaldehyde

Fig 41: Illustration of oxidation.

Reduction:

- Aldehyde/Ketone to Alcohol: $NaBH_4$ (selective), $LiAlH_4$.
- Carboxylic Acid to Alcohol: $LiAlH_4$ ($NaBH_4$ ineffective).
- Ester to Alcohol: $LiAlH_4$ ($NaBH_4$ does not reduce esters).
- Alkene to Alkane: Catalytic hydrogenation.

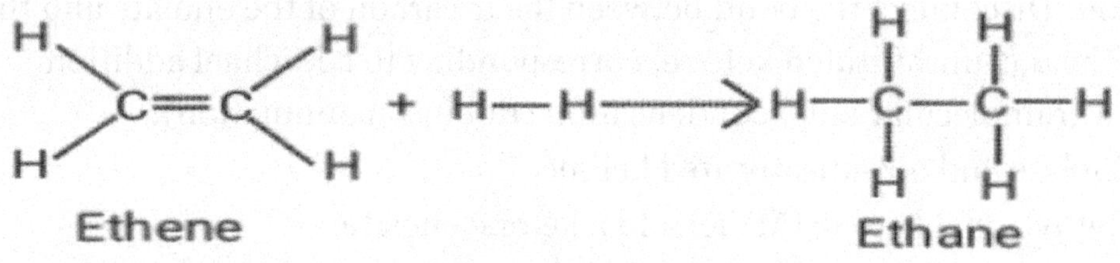

Fig 42: Illustration of reduction.

Hydrolysis:

- Ester to Carboxylic Acid: Acid or base hydrolysis.
- Amide to Carboxylic Acid: Acid or base hydrolysis.

Formation of Carbon-Carbon Bonds:

- Grignard Reaction: Addition of Grignard reagent to carbonyl compounds.
- Wittig Reaction: Reaction of aldehyde or ketone with a Wittig reagent to form an alkene.
- Diels-Alder Reaction: Cycloaddition of a diene and a dienophile to form a cyclohexene.

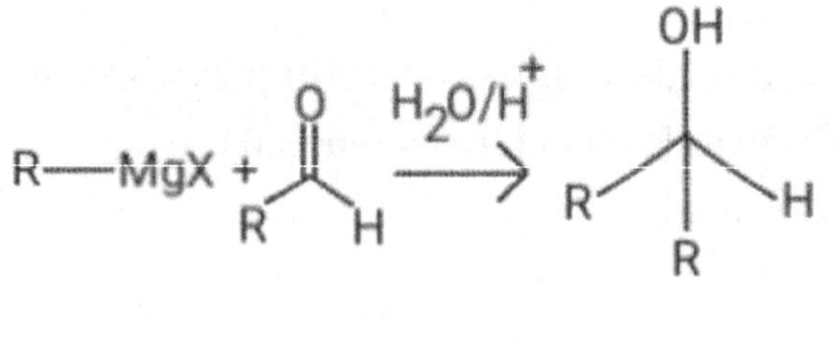

Fig 43: Grignard reaction.

Protection/Deprotection:

- Alcohol Protection: Formation of silyl ethers (e.g., TBSCl, TMSCl).

- Carbonyl Protection: Formation of acetals or ketals.
- Amine Protection: Formation of carbamates (e.g., Boc, Cbz).

Importance of Functional Group Interconversions

- Modify Reactivity: FGIs allow you to introduce or remove functional groups and modify the reactivity of a molecule.
- Protect Reactive Sites: FGIs can be used to protect reactive functional groups that would interfere with the desired transformations.
- Introduce Stereocenters: Some FGIs can be used to introduce stereocenters into a molecule.
- Simplify Synthesis: FGIs can convert complex functional groups into simpler ones.

Protecting Groups

Protecting groups are temporary modifications to a functional group that prevent it from undergoing unwanted reactions during chemical synthesis. The ideal protecting group should be:

- Easy to Introduce: The protecting group should be easy to attach to the functional group.
- Stable to Reaction Conditions: The protecting group should be stable to the reaction conditions used in the synthesis.
- Easy to Remove: The protecting group should be easy to remove under mild conditions and must not affect other functional groups in the molecule.
- Inexpensive: The protecting group should be relatively inexpensive.

Common Protecting Groups

Alcohols

- Silyl Ethers (TMS, TBS, TBDPS): Introduced by reaction with a silyl chloride (e.g., TMSCl, TBSCl, TBDPSCl) in the presence of a base such as imidazole or pyridine and removed by treatment with fluoride sources (e.g., TBAF). Silyl ethers are stable to a wide range of reaction conditions, including acids, bases, and oxidizing agents.

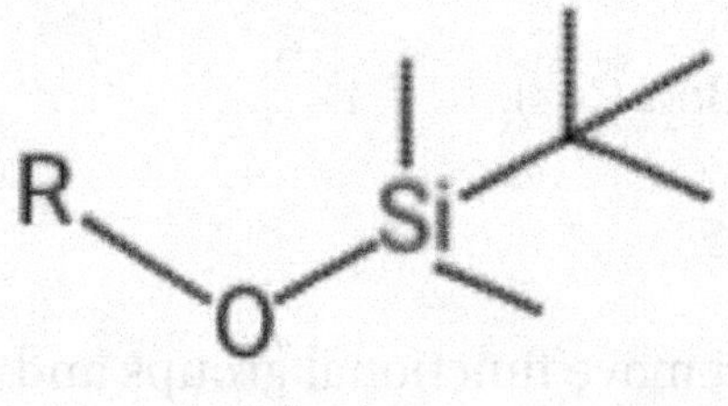

Fig 44: Silyl ethers.

- Benzyl Ethers: Introduced by reaction with benzyl halide (e.g., benzyl bromide) in the presence of a base and removed by catalytic hydrogenation. Benzyl ethers are stable to a wide range of reaction conditions, including acids and bases.

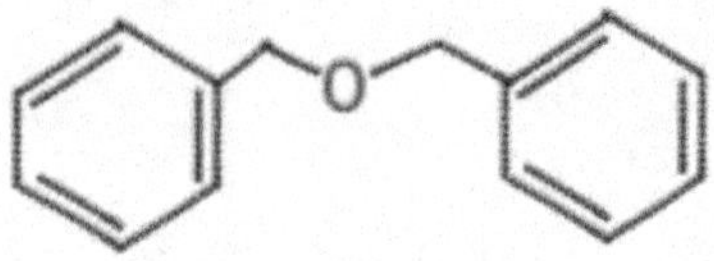

Fig 45: Benzyl ethers.

Carbonyls (Aldehydes and Ketones)

- Acetals and Ketals: Formed by reaction with an alcohol or diol in the presence of an acid catalyst and removed by treatment with aqueous acid. Acetals and ketals are stable to bases, reducing agents, and oxidizing agents.

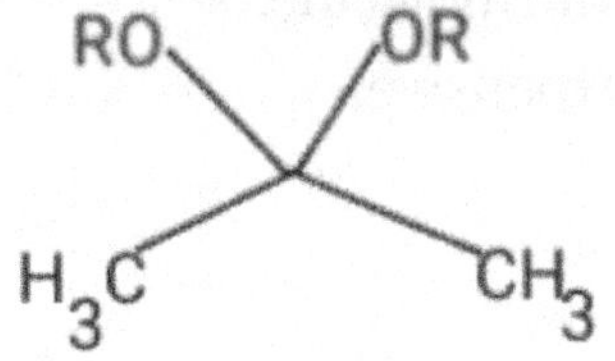

Fig 46: Structure of carbonyls.

Amines

- Carbamates (Boc, Cbz): Boc (tert-butyloxycarbonyl) is introduced by reaction with di-tert-butyl dicarbonate (Boc2O) in the presence of a base and removed by treatment with trifluoroacetic acid or hydrochloric acid (HCl). Cbz (benzyloxycarbonyl) is introduced by reaction with benzyl chloroformate (CbzCl) in the presence of a base and typically removed by catalytic hydrogenation or treatment with strong acids such as HBr in acetic acid. Carbamates are stable to a wide range of reaction conditions, including acids, bases, and oxidizing agents.

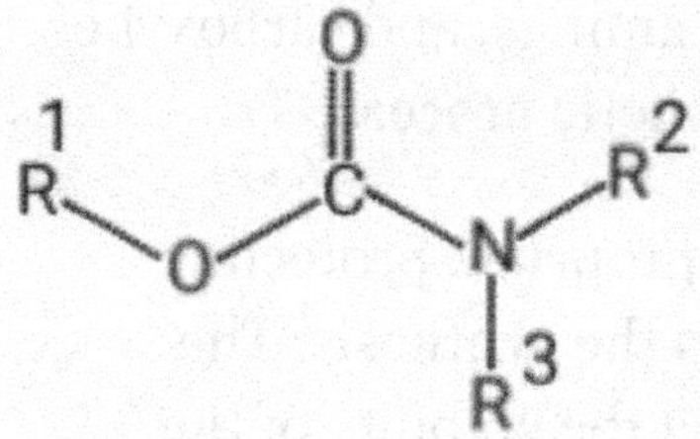

Fig 47: Structure of carbamate functional group.

- Amides (Acetyl, Benzoyl): Introduced by reaction with an acid chloride or anhydride in the presence of a base and removed by treatment with aqueous acid or base. Amides are stable to a wide range of reaction conditions.

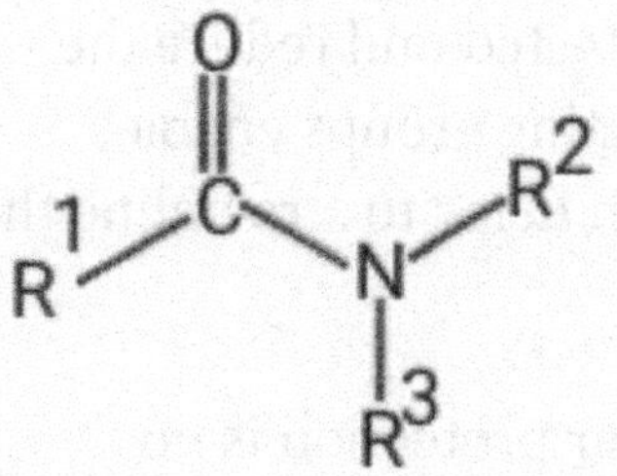

Fig 48: Structure of a general amide.

Carboxylic Acids

- Esters: Can be used to protect carboxylic acids, especially benzyl esters, which can be removed by catalytic hydrogenation.

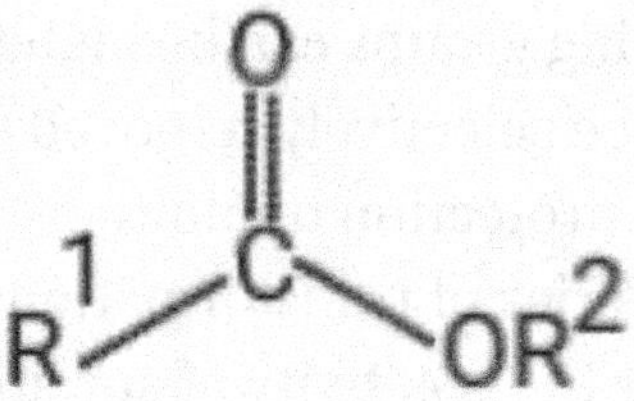

Fig 49: Structure of carboxylic acids.

Strategies for the Use of Protecting Groups

Identify any functional groups within the molecule that may interfere with the desired transformations. Reactive functional groups (such as alcohols, amines, and carboxylic acids) can participate in side reactions that complicate the synthetic process.

Once reactive functional groups have been identified, select appropriate protecting groups that are compatible with the reaction conditions used in the synthesis. The choice of protecting group depends on several factors, including the stability of the group under various reaction conditions, its ease of introduction and removal, and its influence on the overall reactivity of the molecule. Commonly employed protecting groups include tert-butyloxycarbonyl (Boc) for amines and trimethylsilyl (TMS) for alcohols.

To maximize efficiency, protecting groups should be introduced only when necessary in the synthesis to avoid additional steps and complications. Chemists can thus minimize the duration for which reactive functional groups remain unprotected and reduce the risk of unwanted side reactions. The early introduction of protecting groups enables more straightforward management of reaction conditions, which helps to streamline the synthetic pathway.

Conversely, protecting groups should be removed as soon as their protection is no longer required, to avoid unnecessary synthetic complexity. When the removal of protecting groups is delayed, this helps avoid unwanted reactions that could occur if reactive functional groups are exposed too early. When they keep these groups intact until the final stages of synthesis, chemists can maintain the integrity of the desired functional groups and ensure that the final product reflects the intended molecular architecture. This strategy also allows for maximal use of the protecting groups throughout the reaction sequence.

In more complex synthetic routes, the use of orthogonal protecting groups can be highly advantageous. Orthogonal protecting groups are those that can be selectively removed in the presence of one another. They allow for the simultaneous protection of multiple functional groups. This strategy enables chemists to perform a series of transformations on different parts of the molecule and still maintain control over the reactivity of specific functional groups.

Design Multistep Syntheses

A multistep synthesis is a complex and challenging task that requires a deep knowledge of organic reactions, functional group chemistry, and strategic planning. The following guidelines can help you design efficient and effective multistep syntheses.

General Guidelines

- Start with a Retrosynthetic Analysis: Use retrosynthetic analysis to break down the target molecule into simpler starting materials and reaction sequences.
- Consider Functional Group Interconversions: Use FGIs to modify the reactivity and properties of molecules.
- Use Protecting Groups: Use protecting groups to mask reactive functional groups that would interfere with the desired transformations.
- Look for Key Reactions: Identify reactions that can create multiple bonds or stereocenters in a single step.
- Minimize the Number of Steps: Aim for a convergent and efficient synthesis, but balance step count with yield, selectivity, and functional group compatibility to ensure overall practicality.
- Maximize Yields: Choose reactions that give high yields of the desired product.
- Consider Stereochemistry: Carefully consider the stereochemistry of the target molecule and choose reactions that will allow for the controlled introduction of stereocenters.
- Be Practical: Consider the practicality of each step, including the cost of reagents, the availability of starting materials, and the potential for side reactions.
- Plan for Purification: Consider how you will purify the products of each reaction. Choose reactions that give products that are easy to separate from the starting materials and byproducts.
- Be Flexible: Be prepared to modify your synthetic plan if unexpected problems arise.

Advanced Strategies

- Convergent Synthesis: In a convergent synthesis, two or more complex fragments are synthesized separately and then joined together in a final step. This strategy can be more efficient than a linear synthesis in which the molecule is built up step by step from one end.
- Domino Reactions: Domino reactions (also known as cascade reactions or tandem reactions) are sequences of two or more reactions that occur in a single

pot, without the isolation of intermediates. Domino reactions can significantly shorten a synthesis and improve its efficiency.
- Chiral Pool Synthesis: Chiral pool synthesis involves the use of readily available chiral starting materials (e.g., amino acids, carbohydrates, terpenes) to synthesize chiral target molecules. This strategy can be particularly useful for synthesizing natural products.
- Asymmetric Catalysis: Asymmetric catalysis involves the use of chiral catalysts to promote the formation of chiral products in high enantiomeric excess. This strategy is widely used in the synthesis of pharmaceuticals and other fine chemicals.

Examples of Multistep Syntheses

Synthesis of Tamiflu (Oseltamivir)

Tamiflu is an antiviral drug used to treat influenza. Its synthesis involves multiple steps, including:

- A Diels-Alder reaction to form a cyclohexene ring.
- Epoxidation to form an epoxide ring by introducing an oxygen atom across a double bond.
- Azide formation to introduce a nitrogen atom.
- Reductive amination to form an amine.
- Esterification to form the ethyl ester.

Synthesis of Paclitaxel (Taxol)

Paclitaxel is an anticancer drug used to treat a variety of cancers. Its synthesis is extremely complex and involves numerous steps, including:

- Protecting group manipulations.
- Carbon-carbon bond-forming reactions.
- Oxidation and reduction reactions.
- Stereoselective reactions.

Chapter 14: Applications of Organic Chemistry

Organic chemistry is the study of compounds that contain carbon. It is not confined to the laboratory. It permeates nearly every aspect of modern life, from the medicines we take to the materials that surround us.

Organic Chemistry in Pharmaceuticals

The pharmaceutical industry relies heavily on organic chemistry for the discovery, development, and synthesis of new drugs. Organic chemists design and synthesize molecules that interact with specific biological targets, such as enzymes, receptors, and DNA, to treat or prevent diseases.

Drug Discovery

The first step in drug discovery is to identify a biological target that is relevant to the disease of interest. This often involves the study of the molecular mechanisms of the disease and the key proteins or genes that are involved. Once a target is identified, it must be validated to ensure that modulating its activity will have the desired therapeutic effect.

Once a target is validated, the next step is to identify lead compounds that can interact with the target. Lead compounds can be identified through a variety of methods, including:

- High-Throughput Screening (HTS): HTS screens large libraries of compounds against the target to identify those that bind or modulate its activity.
- Fragment-Based Drug Discovery (FBDD): FBDD identifies small, low-affinity fragments that bind to the target, which are then elaborated or linked through medicinal chemistry to create higher-affinity lead compounds.
- Structure-Based Drug Design (SBDD): SBDD uses the three-dimensional structure of the target to design molecules that will bind to it with high affinity and selectivity.
- Natural Products: Natural products, such as plant extracts and microbial metabolites, are a rich source of lead compounds.

Once a lead compound is identified, it must be optimized to improve its potency, selectivity, and pharmacokinetic properties (absorption, distribution, metabolism, and excretion). This often involves synthesizing and testing a series of analogs of the lead compound.

Drug Development

Before a drug can be tested in humans, it must undergo preclinical studies in animals to assess its safety and efficacy. These studies also provide information about the drug's pharmacokinetic properties.

If the preclinical studies are successful, the drug can be tested in humans in clinical trials. Clinical trials are usually conducted in three phases:

- Phase I: Small studies in healthy volunteers to assess the drug's safety, tolerability, pharmacokinetics, and pharmacodynamics.
- Phase II: Larger studies in patients with the disease of interest to assess the drug's efficacy and side effects.
- Phase III: Large, randomized, controlled trials to confirm the drug's efficacy and monitor its long-term safety.

Once a drug has been approved by regulatory agencies (e.g., the Food and Drug Administration in the United States), it must be manufactured on a large scale. Organic chemists play a key role in the development of efficient and cost-effective synthetic routes for drug manufacturing.

Examples of Organic Chemistry in Pharmaceuticals

- Penicillin: A natural product antibiotic discovered by Alexander Fleming. Organic chemists have developed ways to synthesize penicillin and its analogs on a large scale.
- Aspirin: A synthetic drug that is widely used as an analgesic, antipyretic, and anti-inflammatory agent.
- Taxol (Paclitaxel): A natural product that is used to treat a variety of cancers. The synthesis of Taxol is extremely complex and involves numerous steps.
- Ibuprofen: A synthetic nonsteroidal anti-inflammatory drug (NSAID) that is widely used to relieve pain and reduce inflammation.
- Lipitor (Atorvastatin): A synthetic statin drug that is used to lower cholesterol levels.

Organic Materials and Polymers

Organic materials and polymers are ubiquitous in modern life, used in everything from clothes to electronics and aerospace applications. Organic chemistry plays a vital role in the design, synthesis, and characterization of these materials.

Polymers

Polymers are large molecules made up of repeating structural units called monomers. They are classified as either natural polymers (e.g., cellulose, starch, proteins) or synthetic polymers (e.g., polyethylene, polystyrene, nylon).

Polymerization

Polymerization is the process of joining monomers together to form a polymer. There are two main types of polymerizations.

- Addition Polymerization: Monomers add to each other without the loss of any atoms. Examples include the polymerization of ethylene to form polyethylene and the polymerization of styrene to form polystyrene.
- Condensation Polymerization: Monomers join together with the loss of a small molecule, such as water or methanol. Examples include the polymerization of amino acids to form proteins and the polymerization of diacids and diamines to form polyamides (nylons).

Properties of Polymers

The properties of a polymer depend on its chemical structure, molecular weight, and morphology (the arrangement of the polymer chains). Some important properties of polymers include:

- Tensile Strength: The ability of a polymer to resist stretching or breaking.
- Elasticity: The ability of a polymer to return to its original shape after being deformed.
- Glass Transition Temperature (Tg): The temperature at which an amorphous polymer or the amorphous regions of a semi-crystalline polymer transition from a hard, glassy state to a soft, rubbery state.
- Crystallinity: The degree to which the polymer chains are ordered in a crystalline structure. Crystalline polymers tend to be stronger and more rigid than amorphous polymers.

Types of Polymers

- Thermoplastics: Polymers that can be repeatedly softened by heating and solidified by cooling. Examples include polyethylene, polypropylene, polystyrene, and PVC.

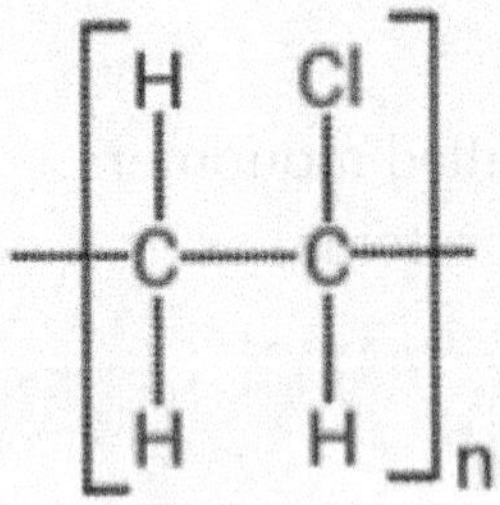

Fig 50: PVC.

- Thermosets: Polymers that undergo irreversible chemical changes during curing (often involving heat or chemical additives) and form a rigid, cross-linked network. Examples include epoxy resins and phenolic resins, but vulcanized rubber is more accurately classified as an elastomer with thermoset-like properties due to sulfur cross-linking.
- Elastomers: Polymers that exhibit rubber-like elasticity. Examples include natural rubber, synthetic rubber (e.g., styrene-butadiene rubber, SBR), and silicone rubber.

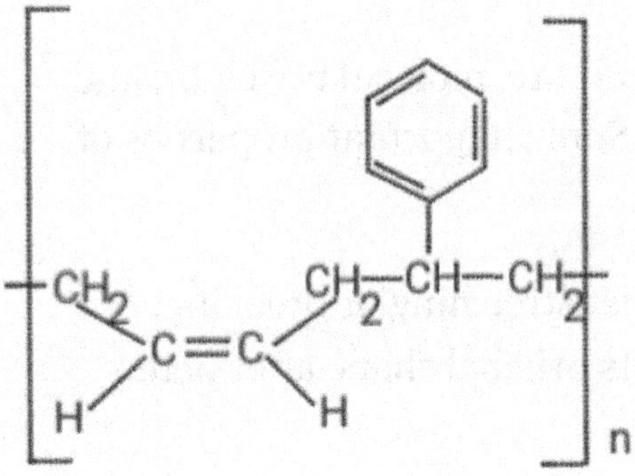

Fig 51: SBR.

Applications of Polymers

Packaging

In the packaging industry, polymers like polyethylene (PE), polypropylene (PP), and polystyrene (PS) dominate due to their low cost and adaptability. Polyethylene, available in low-density and high-density forms, is used for plastic bags, films, and bottles. PP forms rigid containers, such as yogurt containers, and PS creates foam packaging, like Styrofoam. These materials provide good barriers against moisture and gases, but typically offer limited protection against UV light unless stabilized with additives. This extends the shelf life of food, beverages, and other products. Their ability to be molded into various shapes or printed on for branding purposes further enhances their utility.

Biodegradable polymers like polylactic acid (PLA) and polyhydroxyalkanoates (PHA) have become popular as sustainable packaging alternatives, with PLA used in compostable coffee cups. Smart packaging, which incorporates polymer-based sensors or time-temperature indicators, monitors food freshness. Multilayer films combine PE with ethylene vinyl alcohol to improve vacuum-sealed packaging. However, plastic waste remains a significant challenge. This has prompted advancements in recyclable polymers and chemical recycling, which breaks polymers into reusable monomers.

Construction

Polymers play a key role in construction. PVC is a cornerstone for pipes, window frames, siding, and roofing membranes. PVC's durability, resistance to weathering, and low cost make it ideal for long-term use in harsh environments. Polyurethane (PU) foams provide excellent thermal and acoustic insulation, while epoxy resins serve as adhesives and coatings. Compared to metals or concrete, polymers reduce structural weight, which improves efficiency.

New applications include polypropylene-based geotextiles for soil reinforcement in roads and fiber-reinforced polymers like carbon-fiber-reinforced epoxy for bridges. 3D printing with polymers such as acrylonitrile butadiene styrene enables custom construction components. Despite these advancements, fire resistance remains a concern, so the development of flame-retardant additives was necessary. Sustainable construction is also a focus. Self-healing polymers, capable of autonomously repairing cracks, promise to enhance infrastructure longevity.

Textiles

Synthetic polymers like nylon, polyester, and acrylic are staples in textiles. They are used in clothing, carpets, and industrial fabrics due to their strength, elasticity, and resistance to wear. Nylon is found in hosiery and ropes, polyester in apparel and upholstery, and acrylic in sweaters and blankets. These fibers withstand abrasion and repeated washing. They can be engineered for specific properties like moisture-wicking or flame resistance.

The textile industry has expanded into technical and smart applications. Kevlar, an aramid polymer, is used in bulletproof vests, while Gore-Tex (expanded PTFE) creates waterproof, breathable fabrics. Conductive polymers like polyaniline enable smart textiles for wearable electronics, such as heart rate-monitoring sensors. Sustainability is a growing focus. Recycled polyester from PET bottles and bio-based fibers like lyocell reduce environmental impact. However, microplastic pollution from synthetic textiles is a challenge, so research is ongoing into biodegradable fibers and washing machine filtration systems.

Automotive

In the automotive sector, polymers are integral to tires (styrene-butadiene rubber), bumpers (polypropylene), dashboards (ABS), and interior trim (PU). Lightweight composites like carbon-fiber–reinforced polymers are used in high-performance vehicles, which reduces weight to improve fuel efficiency and reduce emissions. Polymers also provide corrosion resistance and design flexibility for aerodynamic shapes.

With the rise of electric vehicles, polymers like polyamide are used in battery housings for their thermal stability. Self-healing polymeric coatings for scratch repair and recycled plastics for sustainability have been developed. High-performance polymers must withstand extreme conditions. This requires advanced thermoplastics and composites. As autonomous vehicles develop, conductive polymers may play a role in sensors and lightweight structures.

Electronics

Polymers are vital in electronics as insulators (polyethylene in cables), semiconductors (conjugated polymers like polythiophene), and dielectrics (polyimide in capacitors). Their flexibility enables bendable electronics like organic light-emitting diode (OLED) screens, while their lightweight nature and low cost make them viable alternatives to silicon. Polyethylene terephthalate serves as a flexible substrate for wearable devices, and organic photovoltaics leverage conjugated polymers like P3HT for lightweight, flexible solar cells.

Flexible electronics, 3D-printed conductive polymer components, and organic electronics are new polymer applications. However, improved conductivity and stability of organic semiconductors remain a challenge. Recyclable electronics to reduce e-waste and polymers for quantum computing components are areas of active research. This highlights the role of polymers in next-generation technologies.

Biomedical

In biomedical applications, polymers are transformative. They enable drug delivery systems, tissue engineering scaffolds, and medical implants. Polylactic-co-glycolic acid facilitates controlled drug release, while polycaprolactone is used in 3D-printed implants for bone regeneration. Biocompatible and biodegradable polymers ensure safe interaction with tissues and harmless degradation in the body.

Innovations include stimuli-responsive polymers for targeted drug delivery and hydrogels for soft robotics and tissue-like implants. Flexible polymers like polydimethylsiloxane are used in wearable biosensors to monitor health metrics. Long-term biocompatibility and minimized immune responses are ongoing challenges. Nanotechnology enhances therapies through nanoparticle drug carriers.

Organic Electronics

Organic electronics is a rapidly growing field that involves the use of organic molecules and polymers as semiconductors in electronic devices. Organic electronic devices provide several advantages over traditional silicon-based devices, which include lower cost, flexibility, and the ability to be printed on flexible substrates.

Organic Light-Emitting Diodes (OLEDs)

OLEDs are a cornerstone of organic electronics, widely used in displays for smartphones, televisions, wearables, and other electronic devices. OLEDs emit light when an electric current passes through organic compounds, eliminating the need for a backlight, unlike traditional liquid crystal displays which require external illumination. This enables thinner, lighter, and more energy-efficient displays with superior contrast, vibrant colors, and faster response times.

The flexibility of organic materials allows OLEDs to be fabricated on bendable or foldable substrates. This enables cutting-edge applications like foldable smartphones and curved television screens. Their low power consumption is particularly advantageous for portable devices and extends battery life. Additionally, OLEDs can be produced via cost-effective techniques like inkjet printing, which supports large-scale manufacturing.

New trends in OLED technology include transparent displays for augmented reality applications and micro-OLEDs for high-resolution virtual reality headsets. Research is also focused on how to improve the longevity of OLED materials, as organic compounds can degrade over time, particularly blue emitters. Sustainable manufacturing, such as using bio-based organic materials, is another area of exploration to reduce environmental impact. Despite these challenges, OLEDs continue to redefine visual technology and provide unmatched display quality and design flexibility.

Organic Photovoltaic Cells (OPVs)

OPVs, or organic solar cells, convert sunlight into electricity using organic semiconductors. Unlike silicon-based solar panels, OPVs are lightweight, flexible, and can be produced at lower costs through roll-to-roll printing processes. These attributes make them ideal for applications where traditional rigid panels are impractical, such as portable chargers, building-integrated photovoltaics, and wearable energy-harvesting devices.

OPVs consist of donor and acceptor organic materials that generate electricity when exposed to light. Their flexibility enables integration into unconventional surfaces, such as windows, fabrics, or curved structures, which expands the potential for solar energy adoption. Additionally, OPVs can be tuned to absorb specific wavelengths. This enables semi-transparent or colored solar cells for aesthetic applications in architecture.

However, OPVs face challenges related to efficiency and stability. Current OPVs achieve power conversion efficiencies that are significantly lower than silicon-based cells, which limits their competitiveness in large-scale energy production. Degradation of organic materials under environmental stress, such as moisture or UV exposure, also affects long-term performance. Ongoing research aims to address these issues through advanced materials, such as non-fullerene acceptors, and encapsulation techniques to enhance durability. Innovations like tandem OPVs, which stack multiple layers to capture more light, have pushed efficiencies closer to commercial viability. As these advancements progress, OPVs hold promise for sustainable, accessible solar energy solutions.

Organic Thin-Film Transistors (OTFTs)

OTFTs are key components in flexible electronic circuits. They serve as switches or amplifiers in applications like flexible displays, wearable sensors, and radio-frequency identification tags. OTFTs utilize organic semiconductors, such as conjugated polymers or small molecules, to conduct charge. They provide a lightweight and bendable alternative to silicon-based transistors.

The main advantage of OTFTs is their compatibility with low-temperature, solution-based processing techniques, such as inkjet or gravure printing. They reduce manufacturing costs and enable production on plastic or paper substrates. This makes OTFTs ideal for large-area electronics, such as flexible e-paper displays or smart packaging with embedded sensors. Their mechanical flexibility also supports wearable devices, like health-monitoring patches that conform to the skin.

Applications of OTFTs have expanded into fields like the Internet of Things, where low-cost, disposable sensors are in demand. For example, OTFT-based circuits can power RFID tags for supply chain tracking or chemical sensors for environmental monitoring. However, OTFTs currently exhibit lower charge mobility than silicon transistors, which limits their use in high-performance computing. Stability under ambient conditions is another challenge, as organic semiconductors can degrade when exposed to air or moisture.

Research is needed to address these limitations, and recent advancements include high-mobility organic materials such as donor-acceptor copolymers, along with improved encapsulation techniques to protect circuits. Hybrid approaches, such as combinations of organic and inorganic materials, balance performance and flexibility. As these advancements unfold, OTFTs are poised to drive the next generation of flexible, low-cost electronics.

Environmental and Green Chemistry

Organic chemistry is important to address environmental challenges and develop sustainable chemical processes. Green chemistry is the design of chemical products and processes that reduce or eliminate the use and generation of hazardous substances.

Principles of Green Chemistry

1) Prevention: It's more effective and environmentally responsible to prevent the creation of waste rather than deal with it after it's been created.
2) Atom Economy: Chemical reactions should be planned so that the maximum proportion of the starting materials are incorporated into the final product, thereby minimizing the generation of waste.
3) Less Hazardous Chemical Syntheses: Wherever possible, reactions should be designed to avoid the use or creation of harmful substances that could pose risks to people or the planet.
4) Designing Safer Chemicals: Products should be engineered to perform their intended role and be as safe and nontoxic as possible.
5) Safer Solvents and Auxiliaries: It's best to avoid solvents or additives if they are not needed—and when they are, to choose options that are safe and environmentally benign.
6) Design for Energy Efficiency: Processes should use as little energy as possible, ideally operating at room temperature and pressure to cut down on environmental and financial costs.

7) Use of Renewable Feedstocks: Raw materials should come from renewable sources (like plants) instead of finite ones whenever it's feasible from both a technical and economic perspective.
8) Reduce Derivatives: Avoid extra steps in chemical processes—such as adding and removing protective groups—when they aren't absolutely necessary since they often generate more waste and require more chemicals.
9) Catalysis: Whenever available, use catalysts because they make reactions more efficient and selective, so large amounts of reagents don't have to be used.
10) Design for Degradation: Chemicals should be made so that, once they're no longer needed, they naturally break down into harmless substances that won't linger in the environment.
11) Real-Time Analysis for Pollution Prevention: Develop and use tools that can monitor chemical reactions as they happen. This helps to avoid the formation of dangerous byproducts.
12) Inherently Safer Chemistry for Accident Prevention: Choose materials and design processes that reduce the chances of spills, fires, or explosions. This makes the work environment and surroundings safer.

Applications of Green Chemistry

Alternative Solvents

One of the cornerstones of green chemistry is the replacement of hazardous organic solvents, such as chloroform or benzene, with safer alternatives like water, $scCO_2$, and ionic liquids. Traditional solvents are often volatile, toxic, and environmentally persistent, and this contributes to pollution and health risks. In contrast, water is nontoxic, abundant, and supports reactions like aqueous-phase catalysis for pharmaceutical synthesis. This reduces the need for harmful organic solvents. $scCO_2$ (supercritical carbon dioxide) exists in a state above its critical temperature and pressure, giving it tunable solvent properties, and it is used in processes like caffeine extraction from coffee beans without leaving toxic residues. Ionic liquids, with their low volatility and recyclability, serve as customizable solvents for reactions like biomass processing.

These alternatives reduce environmental impact and improve worker safety. For example, $scCO_2$ is recyclable, while ionic liquids can be tailored for specific reactions and enhance efficiency. New trends include the development of bio-based solvents, such as those derived from lactic acid, and solvent-free systems like mechanochemistry, where reactions are driven by mechanical force. However, challenges remain, including the high cost of some ionic liquids and the energy required to maintain supercritical conditions. Research aims to optimize these solvents for industrial scalability. This ensures cost-effectiveness and broad adoption.

Catalysis

Catalysis is a pivotal application of green chemistry, which enables reactions to occur under milder conditions, reduces energy use, and minimizes waste. Catalysts accelerate reactions without being consumed. This allows for lower temperatures and pressures compared to traditional methods. For instance, in organic synthesis, catalysts like iron or nickel can serve as more abundant and less toxic alternatives to expensive noble metals like palladium, though palladium is not considered highly toxic but rather costly and less sustainable. Biocatalysis that uses enzymes is particularly promising. It provides high selectivity and operates under ambient conditions. Enzymes are used in industries like pharmaceuticals to produce drugs with fewer byproducts.

Catalysts also enable cleaner processes because they increase reaction specificity and reduce unwanted side products. For example, zeolite catalysts in petrochemical refining improve the yield of desired fuels. Innovations include nanocatalysts, which provide high surface areas for enhanced efficiency, and photocatalysis, where light-driven catalysts degrade pollutants or produce hydrogen. Challenges include catalyst deactivation over time and the need for sustainable catalyst materials. Research is focused on developing recyclable catalysts and bio-inspired systems to address these issues, which makes catalysis a cornerstone of green industrial processes.

Renewable Feedstocks

Green chemistry promotes the use of renewable feedstocks, such as biomass, plant-derived oils, and agricultural waste, over finite petroleum-based resources. Renewable feedstocks reduce reliance on fossil fuels and lower carbon footprints. For example, bioethanol, derived from sugarcane or corn, serves as a renewable platform chemical for the production of plastics like polyethylene. Plant oils such as those from soybeans or castor beans are used to synthesize bio-based lubricants and surfactants. Lignin is a byproduct of paper production and is transformed into valuable chemicals like vanillin through catalytic processes.

These feedstocks support a circular economy because they use waste streams and renewable crops. Advances in biorefineries, which process biomass into fuels, chemicals, and materials, have expanded their applications. For instance, algae-based feedstocks have a high yield and minimal land use. However, challenges include the energy-intensive processing of biomass and competition with food production. Organic chemists continue to develop efficient conversion methods, such as enzymatic hydrolysis, to overcome these barriers. This ensures renewable feedstocks become a viable alternative to petroleum.

Atom-Efficient Reactions

Atom-efficient reactions are a core principle of green chemistry. They aim to maximize the incorporation of all reactants into the final product and minimize waste. Traditional organic reactions often produce significant byproducts, which require costly purification and disposal. Atom-efficient strategies such as click chemistry achieve high yields with minimal waste. The copper-catalyzed azide-alkyne cycloaddition is a click reaction widely used in drug discovery and materials science for its simplicity and efficiency.

Other atom-efficient approaches include cascade reactions, where multiple transformations occur in a single process and reduce the need for intermediate purification. For example, domino reactions in pharmaceutical synthesis streamline the production of complex molecules. These methods lower energy consumption and waste, which aligns with green chemistry's goals. Trends involve atom-efficient reactions combined with renewable feedstocks, such as biomass-derived monomers for polymer synthesis. Challenges include how to design reactions that maintain high efficiency at industrial scales. Advances in computational chemistry aid the design of tailored reaction pathways and enhance the practicality of atom-efficient processes.

Biodegradable Polymers

Biodegradable polymers, designed to be broken down by microorganisms, provide a sustainable alternative to persistent plastics like polyethylene and polystyrene, which contribute to environmental pollution. Green chemistry has driven the development of biopolymers such as PLA (derived from corn starch) and polyhydroxyalkanoates (PHA) (produced by bacteria). These polymers are used in packaging, agricultural films, and biomedical applications like sutures. They are designed to break down into environmentally benign byproducts under appropriate conditions, though complete degradation may depend on specific environmental factors such as temperature, moisture, and microbial activity.

Organic chemistry enables the synthesis of biodegradable polymers with tailored properties, such as flexibility or strength, through copolymerization or blending. For example, PLA is combined with other biopolymers to improve its mechanical performance for food packaging. Innovations include stimuli-responsive polymers that degrade under specific conditions and provide controlled life spans. However, biodegradable polymers often face challenges as they struggle to match the durability and cost of conventional plastics. Research is focused on enhanced performance and reduced production costs. Bio-based monomers from waste streams show promise. These efforts are important to reduce plastic pollution and promote sustainable materials.

Examples of Green Chemistry

- Synthesis of Ibuprofen: A more atom-efficient synthesis of ibuprofen has been developed that reduces the amount of waste generated.
- Synthesis of Biodiesel: Biodiesel is a renewable fuel made from plant-derived oils.
- Enzymatic Synthesis of Pharmaceuticals: Enzymes can be used to catalyze the synthesis of pharmaceuticals under mild conditions and reduce the amount of waste and energy required.

Future Trends and Technologies

Organic chemistry continues to evolve rapidly, and new trends and technologies emerge all the time. Some of the most promising future trends and technologies are mentioned below.

Artificial Intelligence and Machine Learning

Artificial intelligence (AI) and machine learning (ML) have revolutionized organic chemistry. They streamline complex processes like drug discovery, material design, and reaction optimization. AI algorithms analyze vast datasets to predict molecular behavior, identify promising drug candidates, and optimize synthetic routes. For example, ML models can predict the outcomes of chemical reactions and reduce the need for time-consuming experimental trials. In drug discovery, AI platforms like AlphaFold have accelerated the prediction of protein structures, which indirectly supports the identification of protein-ligand interactions and expedites the development of new therapeutics.

The benefits of AI and ML include faster research cycles, reduced costs, and the ability to explore novel chemical spaces. These tools also enhance green chemistry through the design of energy-efficient reactions with minimal waste. Applications include AI-driven retrosynthesis, where algorithms propose synthetic pathways for complex molecules, and autonomous laboratories that use AI to conduct experiments in real time. However, challenges such as the need for high-quality data and the interpretability of AI models persist. Ongoing research is focused on the integration of AI with experimental platforms to create closed-loop systems, which further accelerate innovation in organic chemistry.

Flow Chemistry

Flow chemistry involves chemical reactions conducted in a continuous flow system rather than batch reactors. This has transformed organic synthesis. In flow systems, reactants are pumped through tubes or channels, which allow precise control over reaction parameters like temperature, pressure, and mixing. This results in improved reaction rates, higher yields, and enhanced safety, particularly for hazardous reactions that involve unstable intermediates. For example, flow chemistry is used in pharmaceutical manufacturing to produce active ingredients with consistent quality.

The advantages of flow chemistry include scalability, reduced waste, and the ability to integrate multiple reaction steps in a single system. It aligns with green chemistry principles because it minimizes solvent use and enables real-time monitoring. New trends include the development of modular flow systems for on-demand synthesis and their integration with AI for automated optimization. Challenges include the complexity of flow system design for multistep reactions and the need for specialized equipment. Advances in microfabrication and process engineering continue to address these issues. This positions flow chemistry as a cornerstone of modern chemical manufacturing.

Microreactors

Microreactors are small-scale reactors with channel dimensions in the micrometer range and provide unparalleled control over reaction conditions. This enables the synthesis of complex molecules with high efficiency. Their small volumes facilitate rapid heat and mass transfer, which improves reaction selectivity and reduces side products. Microreactors are used in organic chemistry to synthesize fine chemicals, pharmaceuticals, and nanomaterials, where precision is vital. For instance, they enable the safe handling of exothermic reactions, such as nitrations, that are risky in traditional reactors.

The benefits of microreactors include enhanced safety, reduced reagent consumption, and compatibility with continuous flow systems. They also support green chemistry because they minimize waste and energy use. New applications include point-of-care synthesis of drugs in personalized medicine and the production of high-value chemicals in space-constrained environments. However, challenges such as clogging in microchannels and difficulties in scaling up production remain. Innovations in 3D-printed microreactors and hybrid systems that combine micro- and macro-scale reactors help overcome these barriers. This makes microreactors a versatile tool in organic chemistry.

Biocatalysis

Biocatalysis is the use of enzymes or whole cells to catalyze chemical reactions. It is a sustainable alternative to traditional chemical catalysis. Enzymes provide high selectivity and typically operate under mild conditions (e.g., ambient temperature and physiological or slightly buffered pH), which reduces energy consumption and waste. In organic chemistry, biocatalysis is used to synthesize chiral molecules for pharmaceuticals, such as antibiotics, and produce bio-based chemicals like biofuels. For example, lipases are employed in the synthesis of biodegradable esters for cosmetics.

The environmental benefits of biocatalysis align with green chemistry, as enzymes are biodegradable and derived from renewable sources. Advances in protein engineering, such as directed evolution, have expanded the range of reactions enzymes can catalyze, while synthetic biology enables the design of microbial factories for chemical production. New trends include cascade biocatalysis, where multiple enzymes work sequentially in a single process, and the integration of biocatalysis with flow chemistry for continuous production. Challenges include enzyme stability under industrial conditions and high production costs. Research is focused on immobilizing enzymes for reuse and the development of robust microbial systems. This enhances the scalability of biocatalysis.

Supramolecular Chemistry

Supramolecular chemistry is the study of non-covalent interactions between molecules. It is unlocking new possibilities in organic chemistry by enabling the design of self-assembling systems with tailored functions. These interactions, such as hydrogen bonding and π-π stacking, allow molecules to form complex structures like molecular cages, gels, and host-guest systems. Supramolecular chemistry is used to create smart materials, sensors, and drug delivery systems. For example, cyclodextrin-based supramolecular complexes encapsulate drugs to improve their solubility and targeted release.

The advantages of supramolecular systems include their reversibility (which allows dynamic responses to stimuli like pH or light) and their modularity (which enables rapid design iterations). Applications have expanded into molecular machines, such as synthetic motors for nanotechnology and chemical sensors used to detect pollutants. However, challenges include how to achieve precise control over self-assembly and ensure stability in real-world conditions. Advances in computational modeling and high-resolution imaging aid the design of robust supramolecular systems. This positions this field as a frontier for innovative materials and devices.

Nanomaterials

Organic nanomaterials such as carbon nanotubes, graphene, and conjugated polymers possess unique properties like high conductivity, mechanical strength, and tunable optical characteristics, which make them invaluable in organic chemistry. These materials are used in electronics (e.g., flexible displays), energy storage (e.g., supercapacitors), and biomedicine (e.g., drug delivery nanoparticles). For instance, graphene-based electrodes enhance the performance of organic solar cells, while carbon nanotubes are explored for neural implants due to their biocompatibility.

The versatility of organic nanomaterials stems from their ability to be functionalized with organic molecules. Their properties can be tailored for specific applications. They also support green chemistry because they enable lightweight, energy-efficient devices. New trends include the development of bio-inspired nanomaterials, such as peptide-based nanotubes, and their integration with AI for predictive design. Challenges include the high cost of production, potential toxicity, and difficulties in large-scale synthesis. Research is focused on sustainable synthesis methods, such as biomass-derived nanomaterials, and rigorous safety assessments to ensure environmental and human safety.

Applications of Organic Chemistry in Forensic Science

Forensic science is an interdisciplinary field that integrates principles from various scientific disciplines to address criminal investigations and support the justice system. Among these disciplines, organic chemistry holds a pivotal role due to its focus on carbon-based compounds and their reactions. This branch of chemistry equips forensic scientists with essential tools and techniques necessary to analyze organic materials found at crime scenes. The applications of organic chemistry in forensics are vast. They range from the identification of illicit drugs and toxins to the examination of trace evidence such as fibers, paints, and biological fluids.

Drug Identification and Toxicology

Presumptive and Confirmatory Drug Testing

Organic chemistry is used to identify controlled substances in various forms, including powders, liquids, and biological samples. Drug identification usually involves two stages: presumptive testing and confirmatory testing.

Presumptive Tests

Presumptive tests are designed to provide rapid indications of drug presence, often through observable changes caused by specific organic reactions. These tests are vital in field settings, where quick assessments are necessary.

- Marquis Reagent: This well-known reagent is a mixture of formaldehyde and sulfuric acid that reacts with various alkaloids, such as morphine and heroin, to produce a purple color. When it interacts with amphetamines, the resultant color shifts to an orange-brown hue. The color change arises from acid-catalyzed electrophilic aromatic substitution reactions between the drug and formaldehyde under strongly acidic conditions, making it a valuable preliminary test for drug identification.
- Scott Test: This test is specifically used to detect cocaine. It employs cobalt thiocyanate, which forms a blue complex in the presence of cocaine. The simplicity and speed of this test allow law enforcement to make quick decisions regarding potential drug-related activities.
- Duquenois-Levine Test: This method is widely employed to presumptively identify marijuana. The test produces a purple color upon the addition of Duquenois reagent followed by hydrochloric acid and chloroform, indicating the possible presence of cannabinoids, though it is not specific for THC and may yield false positives.

Despite their speed and utility, presumptive tests lack specificity and can result in false positives. Therefore, confirmatory tests are essential to validate initial findings.

Confirmatory Tests

Confirmatory tests provide precise identification of substances through the use of the unique properties of organic molecules. These methods are usually performed in laboratory settings and provide definitive results.

- Gas Chromatography-Mass Spectrometry (GC-MS): This technique separates vaporized compounds based on their volatility and interactions with the chromatographic column. After separation, the compounds are fragmented to produce a unique mass spectrum, which can be compared against reference libraries. GC-MS is considered the gold standard for drug identification due to its high sensitivity and specificity.
- High-Performance Liquid Chromatography (HPLC): HPLC separates nonvolatile or heat-sensitive drugs through the use of liquid-phase interactions. This method

is particularly useful for substances that cannot withstand the high temperatures required for gas chromatography.
- Fourier Transform Infrared Spectroscopy (FTIR): FTIR measures the absorption of infrared light by organic compounds to detect functional groups. This technique can provide information about specific chemical bonds and molecular structures, which helps identify unknown substances.

Toxicological Analysis

Organic chemistry is used to detect drugs, alcohol, poisons, and metabolites in biological samples such as blood, urine, and hair. Toxicological analysis is important to understand the effects of substances on human health and can provide key evidence in criminal cases.

Analytical Techniques

Various analytical techniques are employed in toxicology to identify and quantify substances in biological samples:

- Immunoassays: These tests utilize antibodies that bind specifically to target molecules. They rely on antigen-antibody interactions and may use enzyme-linked or fluorescent organic molecules for detection, rather than employing organic functional groups for enhanced binding. Immunoassays can detect drugs and their metabolites at very low concentrations and are commonly used for initial screening.
- Liquid Chromatography-Mass Spectrometry (LC-MS): This technique combines the separation capabilities of liquid chromatography with the specificity of mass spectrometry. LC-MS is particularly used to detect drugs and their metabolites in complex biological matrices as it provides accurate and reliable results.
- Gas Chromatography with Nitrogen-Phosphorus Detector (GC-NPD): This method specifically targets nitrogen- or phosphorus-containing compounds, which makes it useful when a variety of drugs and pesticides need to be analyzed. GC-NPD is sensitive and can detect substances at trace levels.

Case Example: Cyanide Poisoning

Organic chemistry is used to detect cyanide, a highly toxic substance. In forensic toxicology, cyanide can be converted into a colored complex, which allows for spectrophotometric analysis. For instance, cyanide can be converted to cyanogen chloride, which then reacts with pyridine and barbituric acid to form a blue-colored complex for spectrophotometric detection in biological samples.

Biological Evidence Analysis

Detection of Blood, Semen, and Saliva

Organic chemistry underpins various tests used to identify biological fluids at crime scenes and provides vital evidence in sexual assault cases and homicides.

Blood Detection

- Kastle-Meyer Test: This classic test employs phenolphthalein, which turns pink when oxidized by hydrogen peroxide in the presence of hemoglobin. The simplicity and effectiveness of this test make it a staple in forensic blood detection.
- Luminol Test: Luminol emits a blue chemiluminescence when it reacts with iron in hemoglobin and reveals trace bloodstains that may not be visible to the naked eye. This test is especially used to detect blood at crime scenes where cleanup efforts may have been made by the criminal.

Semen and Saliva Detection

- Acid Phosphatase Test: This test detects the enzyme acid phosphatase, which is present in high concentrations in semen. The enzyme produces a colored product upon hydrolysis, which makes it a reliable indicator of semen.
- Phadebas Test: The Phadebas test identifies saliva through the detection of the amylase enzyme. When saliva is present, the enzyme breaks down a starch substrate and releases a blue dye that indicates a positive result.

DNA Extraction and Analysis

Organic chemistry is vital for DNA purification and amplification, which are essential in forensic investigations.

- Organic Extraction: This method employs phenol-chloroform to separate DNA from proteins and other cellular materials. The DNA remains in the aqueous phase and allows for its subsequent analysis.
- PCR Reagents: The polymerase chain reaction (PCR) relies on organic molecules, such as nucleotides and primers, to amplify specific DNA sequences. This technique enables forensic scientists to generate sufficient quantities of DNA for analysis, even from small samples.

Fingerprint Analysis

Visualization of Latent Prints

Latent fingerprints are composed of organic residues such as amino acids and fatty acids. They are visualized using techniques rooted in organic chemistry.

Chemical Reagents

- Ninhydrin: This chemical reacts with amino acids present in fingerprints to form Ruhemann's purple, a compound that is used to visualize prints on porous surfaces like paper.
- Cyanoacrylate Fuming: This technique involves polymerizing cyanoacrylate vapors on fingerprint residues and results in the formation of visible white prints. This method is particularly useful for nonporous surfaces.
- Silver Nitrate: Silver nitrate reacts with chloride ions found in sweat to form light-sensitive silver chloride, which can be developed into visible prints under specific lighting conditions.

Fluorescent and Dye Staining

- Rhodamine 6G and DFO: These organic dyes fluoresce under specific light wavelengths and enhance the visibility of fingerprint patterns. Fluorescent techniques can be used to identify latent prints that are otherwise difficult to see.

Recent innovations in fingerprint analysis include the use of luminescent quantum dots and functionalized polymers, which provide improved sensitivity and specificity in fingerprint detection. These advancements enhance the ability of forensic scientists to recover and analyze fingerprints from various surfaces.

Explosives and Arson Investigation

Explosives Detection

Organic explosives, such as TNT and RDX, are identified through the use of a variety of organic chemistry techniques, which play a vital role in bomb investigations.

- GC-MS: This method is employed to separate and identify explosive residues in debris. It provides vital information about the type of explosives used in a detonation.
- HPLC: This allows forensic chemists to identify explosive compounds effectively.

- Infrared Spectroscopy: This technique detects characteristic functional groups, such as nitro groups, which are common in many explosive compounds.

Field Tests

- Griess Test: This test forms a red azo dye when nitrites are present and provides a simple method to detect explosive residues in the field.
- Diphenylamine Test: This method detects oxidizing agents such as nitrates and nitrites and can be used as a preliminary screening tool for explosives.

Arson Investigation

Organic accelerants, such as gasoline and kerosene, are often used in arson cases. Organic chemistry techniques are employed to detect these substances in fire debris.

- Headspace Analysis: This technique extracts volatile compounds from fire debris for subsequent analysis using GC-MS. It allows forensic scientists to identify accelerants that may have been used to ignite a fire.
- Solid Phase Microextraction (SPME): SPME is a sampling technique that concentrates accelerants from complex mixtures, which makes it easier to analyze the organic components present.

Case Example: Gasoline Detection

In arson investigations, GC-MS can identify the unique hydrocarbon pattern associated with gasoline. This enables forensic scientists to distinguish it from other organic sources and confirm its use as an accelerant.

Trace Evidence Analysis

Fibers, Paints, and Polymers

Organic polymers present in fibers, paints, and plastics can be characterized using advanced analytical techniques and provide valuable evidence in forensic investigations.

- Pyrolysis-GC-MS: This method decomposes polymers into smaller fragments and allows forensic scientists to identify the original materials. It is particularly useful for the analysis of synthetic fibers and paints.
- FTIR: Fourier Transform Infrared Spectroscopy is employed to identify functional groups in trace evidence and aids in the characterization of materials found at crime scenes.

- NMR: NMR provides detailed information about the molecular structure of polymers and allows forensic chemists to differentiate between similar materials.

Soil and Glass Analysis

Organic matter in soil, such as humic substances, can be analyzed using GC-MS or HPLC for forensic comparisons. The composition of soil can provide clues about the origin of a suspect or victim.

Similarly, glass evidence can be analyzed to determine its composition and source. This helps link suspects to crime scenes.

Breath Alcohol Testing

Organic chemistry plays a vital role in breathalyzer technology, which is widely used to detect ethanol levels in individuals suspected of driving under the influence.

- Potassium Dichromate Test: This test measures the reduction of orange dichromate ($Cr_2O_7^{2-}$) to blue-green chromium(III) (Cr^{3+}) as ethanol is oxidized. The color change can be measured photometrically to quantify the ethanol concentration.
- Fuel Cell Technology: Modern breathalyzers often utilize fuel cell technology, where the oxidation of ethanol generates an electric current. This current is proportional to the concentration of alcohol in the breath and provides a reliable measurement of blood alcohol content.

Gunshot Residue (GSR) Analysis

GSR is a vital aspect of forensic investigations, particularly in cases that involve firearms. Organic components in GSR, such as nitroglycerin and diphenylamine, can be analyzed with advanced techniques.

- GC-MS: This method is commonly used to analyze GSR and allows forensic scientists to detect and quantify the presence of explosive compounds associated with firearms.
- Liquid Chromatography: Liquid chromatography techniques can also be employed to analyze organic components of GSR, such as stabilizers and plasticizers in propellants, but they are not typically used to determine the specific type of ammunition.

Forensic Entomology

Forensic entomology utilizes organic chemistry to analyze insect-related evidence, which can provide key information in death investigations.

- Cuticular Hydrocarbons: These organic compounds can be used to identify insect species and their developmental stages. Knowledge of the life cycle of insects can help estimate the time of death.
- Volatile Organic Compounds (VOCs): The analysis of VOCs associated with decomposition can help estimate postmortem intervals and provide valuable insights into the timeline of a crime.

Forensic Odontology

Forensic odontology applies organic chemistry principles to the analysis of dental evidence, which can help identify victims or suspects.

- Polymer Analysis: Dental resins and materials can be characterized with techniques such as FTIR and NMR. These analyses help forensic odontologists assess the composition of dental materials.
- Protein Analysis: Analysis of proteins from dental pulp can facilitate DNA extraction. This enables the identification of individuals based on dental records.

Forensic Document Examination

Organic chemistry enables the analysis of ink and paper, which provides insights into document authenticity and alterations.

- Thin-Layer Chromatography (TLC): TLC is used to separate ink dyes for comparison and allows forensic document examiners to identify the composition of inks used in documents.
- FTIR and Raman Spectroscopy: These techniques are employed to identify organic components in inks and papers. This aids in the examination of questioned documents.

Case Example: Forgery Detection

Ink composition analysis can reveal document alterations or estimate the age of ink and provide vital evidence in cases of forgery. The ability to differentiate between inks can help establish the authenticity of documents.

Forensic Anthropology

Forensic anthropology utilizes organic chemistry techniques to analyze skeletal remains and contributes to the identification of individuals in cases of unexplained deaths.

- Amino Acid Racemization: This method measures the conversion of L-amino acids to D-forms over time and is more commonly used to estimate the postmortem interval in ancient or archaeological remains, rather than recent forensic cases.
- Stable Isotope Analysis: Analysis of isotopes in bone collagen can provide insights into an individual's diet and geographic origin, which contributes to the identification process.

Forensic Serology

Forensic serology employs organic chemistry to support blood typing and protein analysis, which can be vital in criminal investigations.

- Organic Antigens and Antibodies: The detection of blood type–specific antigens can help establish connections between suspects and crime scenes.
- Enzyme-Linked Immunosorbent Assay (ELISA): ELISA employs organic molecules in enzyme-substrate reactions to identify specific proteins and enhances the ability to analyze biological evidence.

Environmental Forensics

Environmental forensics investigates environmental crimes and employs organic chemistry techniques to analyze pollutants and contaminants.

- Persistent Organic Pollutants: The analysis of compounds like polychlorinated biphenyls (PCBs) using GC-MS allows forensic scientists to trace sources of contamination and assess environmental damage.
- Oil Spill Analysis: Characterization of the organic composition of oil spills can help identify the source of the spill and determine the extent of environmental impact.

Advances in Forensic Organic Chemistry

Recent advancements in forensic organic chemistry have led to improved detection methods and analytical techniques.

- Nanotechnology: The development of organic sensors, such as organic field-effect transistors and molecularly imprinted polymers, enhances the detection of specific substances in forensic investigations.
- Metabolomics: This new field profiles organic metabolites and provides insights into toxins or drug use. Knowledge of metabolic processes can help identify substances present in biological samples.

Case Studies

- Tylenol Murders (1982): In this notorious case, organic extraction and colorimetric assays were employed to identify cyanide in capsules. This prompted significant regulatory changes in the pharmaceutical industry.
- Unabomber Investigation: The use of GC-MS and infrared spectroscopy linked organic explosive residues to the suspect's materials. This provided vital evidence in a high-profile case.
- O.J. Simpson Trial: Organic chemistry contributed to the analysis of blood evidence, but the detection of EDTA, a preservative found in blood collection tubes, was highly controversial and its forensic interpretation was debated during the trial.

Test 1: Questions

(1) What is the purpose of the IUPAC nomenclature system in the context of organic chemistry?

(A) Create distinctive abbreviations for various organic compounds.

(B) Provide a systematic and unambiguous name for every organic compound.

(C) Classify organic compounds based on their physical properties.

(D) Simplify the process of naming inorganic compounds.

(2) Which specific component of an IUPAC name indicates the location and name of attached groups or substituents?

(A) The stereochemistry identifier.

(B) The locant/substituents.

(C) The parent chain.

(D) The suffix.

(3) In the IUPAC name 3-methylpent-2-ene, what does "pent" signify regarding the structure of the compound?

(A) A carbon chain consisting of five carbon atoms.

(B) The presence of a triple bond within the structure.

(C) A methyl substituent attached to the main chain.

(D) An alcohol functional group within the compound.

(4) What significance does the suffix 2-ene have in the context of the name 3-methylpent-2-ene?

(A) It indicates that there is a double bond beginning at carbon 2.

(B) It signifies the presence of a triple bond at carbon 2.

(C) It suggests that there are two methyl groups attached.

(D) It denotes the existence of an alcohol group situated at carbon 2.

(5) What is the initial step when applying IUPAC rules to name an organic compound?

(A) Number the parent chain of the compound.

(B) Identify the principal functional group present in the molecule.

(C) List all substituents in alphabetical order.

(D) Combine prefixes and suffixes to create the complete name.

(6) What fundamental concept does valence bond theory elucidate through the idea of orbital overlap?

(A) The nature of molecular polarity, which relates to charge distribution.

(B) The process of chemical bonding between atoms.

(C) The interactions that occur between different molecules.

(D) The properties observed during spectroscopic analysis.

(7) What is the geometric arrangement of a carbon atom that exhibits sp^3 hybridization?

(A) Linear.

(B) Trigonal planar.

(C) Tetrahedral.

(D) Trigonal bipyramidal.

(8) Which of the following molecules serves as an example of sp^2 hybridization?

(A) Propyne (C_3H_4).

(B) Propene (C_3H_6).

(C) Propane (C_3H_8).

(D) Hydrogen cyanide (HCN).

(9) What is the bond angle associated with sp hybridization?

(A) 109.5°.

(B) 120°.

(C) 180°.

(D) 90°.

(10) Why are the C–H bonds in sp-hybridized carbon atoms stronger than those in sp^3-hybridized carbon atoms?

(A) Because they have greater p-character in their hybrid orbitals.

(B) Due to longer bond lengths associated with the bonds.

(C) Because of the higher s-character in sp-hybridized orbitals.

(D) As a result of having more π bonds present.

(11) Which of the following statements regarding the Arrhenius definition of acids and bases during the year 1884 is correct?

(A) Acids are substances that produce H^+ ions, and bases are substances that produce OH^- ions in water.

(B) Acids are substances that produce OH^- ions, and bases are substances that produce H^+ ions in water.

(C) Acids are substances that produce H^+ ions, and bases are substances that produce H_3O^+ ions in water.

(D) Acids are substances that produce OH^- ions, and bases are substances that produce H_3O^+ ions in water.

(12) Which ion is most commonly used to represent the H^+ ion in aqueous solutions according to the Arrhenius definition?

(A) Hydroxide ion (OH^-).

(B) Hydronium ion (H_3O^+).

(C) Zundel cation ($H_5O_2^+$).

(D) Eigen cation ($H_9O_4^+$).

(13) Which of the following statements best represents an acid according to the Brønsted-Lowry theory?

(A) A molecule that donates a proton (H^+).

(B) A molecule that accepts a proton (H^+).

(C) A species capable of accepting an electron pair.

(D) A species capable of donating an electron pair.

(14) Which of the following statements best represents a Lewis acid according to G.N. Lewis?

(A) A species capable of donating an electron pair.

(B) A species capable of accepting an electron pair.

(C) A species that produces H^+ ions.

(D) A species that produces OH^- ions.

(15) All of the following are examples of a Lewis acid except:

(A) $AlCl_3$.

(B) BF_3.

(C) $FeCl_3$.

(D) HCl.

(16) How do skeletal (chain) isomers differ from other types of constitutional isomers?

(A) The positions of substituents.

(B) The functional groups present.

(C) The branching of their carbon skeletons.

(D) The stereochemistry of the molecule.

(17) How does the increased branching in isopentane affect its boiling point compared to its skeletal isomer n-pentane?

(A) Increases the boiling point.

(B) Decreases the boiling point.

(C) Has no effect on the boiling point.

(D) Increases intermolecular forces.

(18) How do the physical properties of skeletal isomers relate to their molecular structures?

(A) They are only determined by molecular weight.

(B) They are only determined by the presence of functional groups.

(C) They vary due to differences in branching and surface area.

(D) They remain constant regardless of structural differences.

(19) How do positional isomers differ from skeletal isomers?

(A) Positional isomers have different carbon skeletons.

(B) Positional isomers vary in the location of functional groups.

(C) Skeletal isomers contain different functional groups.

(D) Skeletal isomers have the same carbon backbone.

(20) Which of the following distinguishes functional isomers from other types of isomers?

(A) Differences in carbon backbone.

(B) Varied locations of functional groups.

(C) Presence of different functional groups.

(D) Varying stereochemistry.

(21) What type of reaction replaces a leaving group with a nucleophile?

(A) Elimination.

(B) Nucleophilic substitution.

(C) Addition.

(D) Rearrangement.

(22) Which of the following is the correct representation of the rate law for an SN2 reaction?

(A) Rate = k[substrate].

(B) Rate = k[nucleophile].

(C) Rate = k[substrate][nucleophile].

(D) Rate = $k[\text{substrate}]^2$.

(23) Which type of substrates does the SN2 mechanism favor?

(A) Tertiary substrates.

(B) Secondary substrates only.

(C) Primary and unhindered secondary alkyl halides.

(D) All types of substrates equally.

(24) Which of the following statements regarding the effect of polar aprotic solvents on the SN2 reaction is correct?

(A) They hinder nucleophilicity.

(B) They make anionic nucleophiles less reactive.

(C) They increase the reactivity of nucleophiles.

(D) They are not suitable for SN2 reactions.

(25) All of the following are regarded as good leaving groups in SN2 reactions except:

(A) Hydroxide ion (OH^-).

(B) Protonated water.

(C) Halides (I^-, Br^-, Cl^-).

(D) Mesylate (OMs^-).

(26) What type of bonds allow alkenes and alkynes to react with electrophiles?

(A) σ bonds.

(B) π bonds.

(C) δ bonds.

(D) ϕ bonds.

(27) Which of the following products is formed during the electrophilic attack stage of alkenes and alkynes?

(A) A stable alkene.

(B) An alkane.

(C) Addition product.

(D) A carbocation intermediate.

(28) Which of the following products is formed during the nucleophilic attack stage of alkenes and alkynes?

(A) A stable alkene.

(B) An alkane.

(C) Addition product.

(D) A carbocation intermediate.

(29) What is the chief purpose of spectroscopy in organic chemistry?

(A) Synthesize new compounds.

(B) Determine molecular structures.

(C) Predict chemical reactions.

(D) Measure pH levels.

(30) What type of radiation does infrared (IR) spectroscopy utilize?

(A) Ultraviolet radiation.

(B) Visible light.

(C) Infrared radiation.

(D) X-rays.

(31) Which of the following phenomena happens when the frequency of IR radiation matches a bond's vibrational frequency?

(A) The bond absorbs energy, and the vibrational amplitude increases.

(B) The bond breaks.

(C) The bond emits radiation.

(D) The molecule becomes ionized.

(32) What is the correct definition of wavenumber in the context of IR spectroscopy?

(A) The number of wavelengths per unit distance.

(B) The number of molecules in a sample.

(C) The energy of a photon.

(D) The speed of sound in a medium.

(33) Which region of the IR spectrum is unique to each molecule and used for identification?

(A) Functional group region.

(B) Absorption region.

(C) Fingerprint region.

(D) Vibrational region.

(34) What is the relationship between wavenumber and frequency?

(A) Wavenumber is inversely proportional to frequency.

(B) Wavenumber is unrelated to frequency.

(C) Wavenumber is directly proportional to frequency.

(D) Wavenumber equals frequency.

(35) Which region of the IR spectrum generally contains more distinct peaks for functional groups?

(A) Below 1,500 cm^{-1}.

(B) Above 1,500 cm^{-1}.

(C) Around 1,000 cm^{-1}.

(D) Below 1,000 cm^{-1}.

(36) Which of the following correctly states the three main stages of radical reactions?

(A) Initiation, propagation, termination.

(B) Initiation, termination, propagation.

(C) Propagation, termination, initiation.

(D) Generation, combination, termination.

(37) Which factor is the most common trigger in the initiation stage of radical reactions?

(A) Combination of radicals.

(B) Generation of stable molecules.

(C) Homolytic cleavage of bonds.

(D) Absorption of heat only.

(38) How does the termination stage affect the radical chain reaction?

(A) It increases the number of radicals.

(B) It removes radicals and halts the reaction.

(C) It generates new radicals.

(D) It initiates further radical formation.

(39) What occurs during the propagation stage of radical chain reaction?

(A) Radicals are generated.

(B) Radicals react with stable molecules to form new radicals.

(C) Radicals combine to stop the reaction.

(D) Stable molecules are transformed into radicals.

(40) What happens to the concentration of radicals during the propagation stage of the radical chain reaction?

(A) It increases significantly.

(B) It decreases rapidly.

(C) It fluctuates wildly.

(D) It remains constant.

(41) Which of the following statements correctly characterizes aromatic compounds?

(A) They have unique stability due to delocalized π electrons.

(B) They are highly reactive.

(C) They have a linear structure.

(D) They lack a cyclic structure.

(42) What happens during the deprotonation step of electrophilic aromatic substitution (EAS)?

(A) A proton is added to the aromatic ring.

(B) Aromaticity is lost.

(C) A base abstracts a proton and restores aromaticity.

(D) The electrophile is removed.

(43) What is the driving force for the deprotonation step in EAS?

(A) Formation of a stable carbocation.

(B) Regeneration of the electrophile.

(C) Restoration of aromaticity.

(D) Addition of hydrogen gas.

(44) What type of reactions do aromatic compounds usually undergo?

(A) Addition reactions.

(B) Substitution reactions.

(C) Elimination reactions.

(D) Rearrangement reactions.

(45) What is the first step in the mechanism of EAS?

(A) Deprotonation.

(B) Electrophilic attack.

(C) Formation of a stable aromatic compound.

(D) Formation of a carbocation.

(46) Which of the following is the correct general formula for aldehydes?

(A) RCOR'.

(B) RCHO.

(C) RCH_2OH.

(D) $R_2C=O$.

(47) Which of the following statements about the carbonyl group is true?

(A) It is nonpolar.

(B) It has a partial positive charge on oxygen.

(C) It has a partial positive charge on carbon.

(D) It cannot interact with nucleophiles.

(48) What effect does increasing the size of alkyl groups have on the solubility of aldehydes and ketones in water?

(A) Solubility decreases.

(B) Solubility increases.

(C) Solubility remains constant.

(D) Solubility becomes unpredictable.

(49) What type of interaction contributes to the higher boiling points of aldehydes and ketones compared to alkanes?

(A) Hydrogen bonding.

(B) Ionic interactions.

(C) London dispersion forces.

(D) Dipole-dipole interactions.

(50) Why do aldehydes and ketones generally have lower boiling points than alcohols of comparable molecular weight?

(A) They cannot form hydrogen bonds with each other.

(B) They are nonpolar.

(C) They have higher molecular weights.

(D) They form stronger ionic interactions.

(51) What type of isomerism does keto-enol tautomerism represent?

(A) Geometric isomerism.

(B) Structural isomerism.

(C) Stereoisomerism.

(D) Enantiomerism.

(52) Which of the following is the correct first step of the mechanism in acid-catalyzed tautomerism?

(A) Protonation of the carbonyl oxygen.

(B) Deprotonation of the α-carbon.

(C) Formation of the enolate ion.

(D) Migration of a double bond.

(53) Which of the following is the correct structure of the enol form?

(A) A carbonyl group with an adjacent hydroxyl group.

(B) A carbonyl group with a single bond to an alcohol.

(C) An alcohol with a double bond adjacent to it.

(D) A saturated hydrocarbon.

(54) Which of the following statements regarding keto-enol tautomerism is true?

(A) The keto form is more stable than the enol form in most cases.

(B) The enol form is more stable than the keto form in most cases.

(C) The keto and enol forms cannot interconvert.

(D) Both forms exist in a fixed ratio without any interconversion.

(55) What effect does the solvent have on the equilibrium between keto and enol forms?

(A) It does not affect the equilibrium.

(B) It can shift the equilibrium toward the keto or enol form depending on polarity.

(C) It only favors the formation of the enol form.

(D) It prevents any tautomerization from occurring.

(56) Which statement regarding oxidation in the context of organic chemistry is correct?

(A) Loss of electrons or increase in oxidation state.

(B) Decrease in oxidation state.

(C) Formation of carbonyl compounds.

(D) Gain of electrons.

(57) Which of the following is the first step in the mechanism of oxidation with PCC?

(A) Elimination of a chromium (IV) species.

(B) Formation of a chromate ester.

(C) Protonation of the alcohol.

(D) Reduction to a carbonyl compound.

(58) What is a significant advantage of using PCC as an oxidizing agent?

(A) It over-oxidizes aldehydes to carboxylic acids.

(B) It is a potent oxidizing agent.

(C) It is mild and selective.

(D) It requires no solvent.

(59) Which of the following statements regarding the limitation of PCC is correct?

(A) It is ineffective in oxidizing alcohols.

(B) It is not toxic at all.

(C) It is only effective in aqueous solutions.

(D) It is a stoichiometric reagent.

(60) What type of alcohols does CrO_3 oxidize to carboxylic acids?

(A) Tertiary alcohols.

(B) Secondary alcohols.

(C) Primary alcohols.

(D) All types of alcohols.

(61) Which of the following sentences correctly describes disconnection in the context of retrosynthetic analysis?

(A) The completion of a synthesis.

(B) The mental breaking of a bond in the target molecule.

(C) The isolation of a product.

(D) The reaction of two compounds.

(62) Which of the following sentences correctly describes a synthon in the context of retrosynthetic analysis?

(A) A stable molecule used in synthesis.

(B) A reagent that cannot be used in reactions.

(C) A product of a chemical reaction.

(D) An idealized fragment of a molecule that results from a disconnection.

(63) Which of the following is the correct first step in retrosynthetic analysis?

(A) Analyze the target molecule.

(B) Identify reagents and reactions.

(C) Indicate the synthons.

(D) Write the forward synthesis.

(64) What key elements should be identified when analyzing the target molecule?

(A) Functional groups, stereocenters, and structural features.

(B) Molecular weight and boiling point.

(C) Solubility and density.

(D) Color and odor.

(65) What is the importance of disconnecting bonds adjacent to functional groups?

(A) It always simplifies the synthesis.

(B) It prevents unwanted side reactions.

(C) These locations are usually sites of reactivity.

(D) It makes the target molecule more complex.

(66) Which of the following should be conducted as a first step in drug discovery?

(A) Optimize lead compounds.

(B) Validate the target.

(C) Identify a biological target.

(D) Conduct clinical trials.

(67) How does high-throughput screening (HTS) identify lead compounds?

(A) Test small libraries of compounds against a target.

(B) Identify small fragments that bind to the target and link them.

(C) Analyze the structure of lead compounds.

(D) Design molecules based on the three-dimensional structure of the target.

(68) What function does fragment-based drug discovery (FBDD) perform in the context of lead compound identification?

(A) Identifies small fragments that bind to the target and links them.

(B) Designs molecules based on the three-dimensional structure of the target.

(C) Tests small libraries of compounds against a target.

(D) Designs drugs based on the target's chemical formula.

(69) What does ADME stand for in pharmacokinetics?

(A) Absorption, distribution, metabolism, excretion.

(B) Application, development, modification, evaluation.

(C) Analysis, design, manufacturing, efficacy.

(D) Active, dynamic, molecular, environmental.

(70) Which of the following assessments must be conducted before a drug can be tested in humans?

(A) Clinical trials.

(B) Regulatory approval.

(C) Preclinical studies.

(D) Large-scale manufacturing tests.

Test 1: Answers and Explanations

(1) (B) Provide a systematic and unambiguous name for every organic compound.

The IUPAC nomenclature system is designed to ensure that every organic compound has a systematic and precise name. It facilitates effective communication among chemists globally. The creation of abbreviations is not its chief goal, and classification based on physical properties does not align with its purpose. IUPAC nomenclature mainly applies to organic compounds rather than inorganic ones.

(2) (B) The locant/substituents.

The locant/substituents segment of an IUPAC name provides both the positions (locants) and the names of groups that are attached to the main carbon chain, such as 3-methyl in 3-methylpent-2-ene. The stereochemistry identifier relates to the spatial arrangement of atoms and does not address substituents. The parent chain identifies the longest carbon chain, while the suffix indicates the functional group but does not specify attached groups.

(3) (A) A carbon chain consisting of five carbon atoms.

The suffix "pent" refers specifically to a parent chain that contains five carbon atoms, which follows the established IUPAC conventions for naming hydrocarbons. The indication of a triple bond would require a different suffix, while "methyl" refers to a substituent and not the root name itself. Additionally, the name does not contain any indication of an alcohol group.

(4) (A) It indicates that there is a double bond beginning at carbon 2.

The suffix -2-ene indicates that there is a double bond starting at carbon 2 in the hydrocarbon chain. If a triple bond were present, a different suffix would be used. The suffix does not imply the presence of multiple methyl groups or an alcohol functional group, which would be designated differently.

(5) (B) Identify the principal functional group present in the molecule.

The first step in the IUPAC naming process is to identify the principal functional group. This choice determines the suffix of the name and informs subsequent decisions regarding chain selection and numbering. The protocol numbers the parent chain after identifying the functional group. The procedure lists substituents and assembles the complete name later in the process.

(6) (B) The process of chemical bonding between atoms.

Valence bond theory provides a comprehensive explanation of chemical bonding. It highlights how atomic orbitals overlap to form bonds, such as sigma (σ) bonds found in organic molecules. The concept of molecular polarity pertains to properties rather than the mechanics of bonding itself. Intermolecular forces concern interactions between molecules. Spectroscopic properties relate to light absorption and emission, which are unrelated to the theoretical framework of bonding.

(7) (C) Tetrahedral.

When a carbon atom undergoes sp^3 hybridization, it forms four equivalent orbitals. This action results in a tetrahedral geometry with bond angles of approximately 109.5°, as seen in propane (C_3H_8). A linear arrangement corresponds to sp, trigonal planar relates to sp^2, and trigonal bipyramidal is not usual for carbon.

(8) (B) Propene (C_3H_6).

Propene (C_3H_6) features a carbon atom with sp^2 hybridization due to the presence of a double bond, which results in a trigonal planar arrangement with bond angles of 120°. Propyne has sp-hybridized carbons because of the triple bond, propane consists entirely of sp^3 hybridization with single bonds, and hydrogen cyanide contains sp-hybridized carbon.

(9) (C) 180°.

In sp hybridization, two linear orbitals are formed, which results in a bond angle of 180°, as exemplified in prop-1-yne (C_3H_4). The angles 109.5° and 120° correspond to sp^3 and sp^2 hybridization, respectively, while 90° does not apply to any common hybridization states for carbon.

(10) (C) Because of the higher s-character in sp-hybridized orbitals.

The sp-hybridized orbitals possess 50% s-character, which draws electrons closer to the nucleus. This action results in shorter and stronger bonds compared to sp^3 hybridized bonds, which have only 25% s-character. Sp hybridization has less p-character than sp^2 or sp^3. Sp bonds are indeed shorter. C–H bonds are σ bonds, not π bonds.

(11) (A) Acids are substances that produce H^+ ions, and bases are substances that produce OH^- ions in water.

Acids and bases play a pivotal role in organic reaction mechanisms. They affect everything from reaction rates to product selectivity. Arrhenius introduced one of the first definitions of acids and bases in 1884, which states that acids are substances that produce H^+ ions in water, while bases produce OH^- ions.

(12) (B) Hydronium ion (H_3O^+).

The Arrhenius definition needs to be revised to reflect current understanding: in aqueous solutions, H^+ does not exist freely but rather as a mixture of hydronium (H_3O^+), the Zundel cation ($H_5O_2^+$), and the Eigen cation ($H_9O_4^+$). Usually, only the hydronium ion is represented. Although the Arrhenius definition is helpful for water-based scenarios, it is limited and does not encompass the entire range of acid-base chemistry.

(13) (A) A molecule that donates a proton (H^+).

In the Brønsted-Lowry acid-base theory established in 1923, an acid is defined as a molecule that donates a proton (H^+), while a base accepts a proton.

Example: $HCl + H_2O \rightleftharpoons Cl^- + H_3O^+$

In this reaction, HCl donates a proton to water and produces hydronium (H_3O^+). Water acts as the base and accepts a proton from HCl, resulting in the formation of chloride (Cl^-, the conjugate base) and hydronium (H_3O^+, the conjugate acid).

(14) (B) A species capable of accepting an electron pair.

The Lewis definition expands our understanding of acids and bases beyond the traditional Brønsted-Lowry (focused on proton transfer) and Arrhenius (involving H^+/OH^- in water) definitions. According to G.N. Lewis, a Lewis acid is a species capable of accepting an electron pair, which makes it an electron pair acceptor.

(15) (D) HCl.

Lewis acids are often electron-deficient and feature an incomplete octet or a positive charge. Examples include BF_3, $AlCl_3$, $FeCl_3$, and H^+. Conversely, a Lewis base is a species that donates an electron pair and acts as an electron pair donor. Lewis bases usually possess lone pairs of electrons. Examples include NH_3, H_2O, OH^-, and species with π bonds that can donate electrons.

(16) (C) The branching of their carbon skeletons.

Skeletal isomers differ in the branching of their carbon skeletons. These variations can significantly affect the properties of the compounds. For example, consider the two isomers of pentane:

- n-Pentane has a straight-chain structure with five carbon atoms connected linearly (C_5H_{12}).
- Isopentane (2-methylbutane) is a branched isomer that has a carbon chain with a branching point, which results in a different structural arrangement (C_5H_{12}).
- Neopentane (2,2-dimethylpropane) is a central carbon atom bonded to three methyl groups and one hydrogen ($C(CH_3)_4$).

(17) (B) Decreases the boiling point.

n-Pentane boils at approximately 36.1°C. Its linear structure allows for greater surface contact between molecules, which strengthens London dispersion forces. Isopentane boils at approximately 27.8°C. The branching reduces surface area, which weakens intermolecular forces. Neopentane boils at approximately 9.5°C. Its highly compact structure minimizes intermolecular interactions, which results in the lowest boiling point.

(18) (C) They vary due to differences in branching and surface area.

The difference in branching results in distinct boiling points and densities. These differences demonstrate how skeletal isomers can exhibit varying physical properties despite having the same molecular formula. Boiling points are largely determined by the strength of intermolecular forces, which depend on molecular shape. Linear molecules have stronger van der Waals forces due to increased surface area, while branched molecules are more compact, which reduces these forces.

(19) (B) Positional isomers vary in the location of functional groups.

Positional isomers vary in the location of functional groups within the carbon chain. Unlike skeletal isomers, which involve differences in the carbon backbone, positional isomers maintain the same carbon framework but vary in where the functional group or substituent is located.

(20) (C) Presence of different functional groups.

Functional isomers contain different functional groups, which can cause distinct chemical properties and reactivity profiles. This type of isomerism is distinct from skeletal or positional isomerism. It involves a fundamental change in the type of chemical bonding or functional group rather than the arrangement of the carbon skeleton or the position of a functional group.

(21) (B) Nucleophilic substitution.

Nucleophilic substitution and elimination reactions either replace a leaving group with a nucleophile or remove both a leaving group and a hydrogen atom to form a π bond. Each reaction type can proceed through two main mechanisms: unimolecular (SN1 and E1) and bimolecular (SN2 and E2).

(22) (C) Rate = k[substrate][nucleophile].

The reaction rate depends on both the substrate and nucleophile concentrations. Therefore, the rate law of SN2 is second order.

(23) (C) Primary and unhindered secondary alkyl halides.

Primary and unhindered secondary alkyl halides favor SN2. Tertiary substrates are generally unreactive due to steric hindrance. Moreover, a strong nucleophile is required, usually an anionic or a neutral nucleophile with a lone pair.

(24) (C) They increase the reactivity of nucleophiles.

Polar aprotic solvents (e.g., acetone, DMSO, DMF) are preferred as they solvate cations without significantly solvating anionic nucleophiles, which makes them more reactive. Polar protic solvents can hinder nucleophilicity by forming hydrogen bonds.

(25) (A) Hydroxide ion (OH^-).

A good leaving group is essential (weak base, stable after departure). Common good leaving groups include halides ($I^- > Br^- > Cl^-$), tosylate (OTs^-), mesylate (OMs^-), and protonated water. In SN2 reactions, a poor leaving group is a strong base and is not readily displaced by the nucleophile. This is because strong bases have a high affinity for electrons and are, therefore, more stable in their anionic form. Examples of poor leaving groups include hydroxide (OH^-), alkoxide (RO^-), amide ($NH2^-$), and alkyl (R^-) ions.

(26) (B) π bonds.

Alkenes and alkynes are characterized by their π bonds (areas of high electron density). They readily undergo attacks by electrophiles (electron-seeking species). During this process, the π bond breaks. This results in the formation of new σ bonds with the electrophile and another atom or group.

(27) (D) A carbocation intermediate.

During an electrophilic attack, the electrophile (E^+) approaches the electron-rich π bond, which forms a bond with one of the carbon atoms involved in the multiple bond. This step consists of the donation of π electrons to the electrophile, which results in a carbocation intermediate on the other carbon of the multiple bond.

(28) (C) Addition product.

In a nucleophilic attack, the carbocation, being electron-deficient, is subsequently attacked by a nucleophile (Nu^- or a neutral species with a lone pair). This results in the formation of the addition product.

(29) (B) Determine molecular structures.

Spectroscopy is a technique in organic chemistry used to determine molecular structures. It enables us to visualize molecules at an atomic level by examining their interactions with various forms of electromagnetic radiation and analyzing how they fragment.

(30) (C) Infrared radiation.

Infrared (IR) spectroscopy relies on the principle that molecules absorb specific frequencies of infrared radiation that correspond to the vibrational frequencies of their bonds.

(31) (A) The bond absorbs energy, and the vibrational amplitude increases.

When the frequency of IR radiation matches a bond's vibrational frequency, the bond absorbs energy, which increases its vibrational amplitude. We analyze the absorbed frequencies to identify the functional groups present in a molecule.

(32) (A) The number of wavelengths per unit distance.

Wavenumber ($\tilde{v}$) is defined as the number of wavelengths per unit distance and is expressed in inverse centimeters (cm^{-1}). It is directly proportional to frequency ($\tilde{v} = c/\tilde{v}$, where c is the speed of light) and energy ($E = hv = hc/\tilde{v}$, where h is Planck's constant). IR spectra are usually displayed as transmittance or absorbance against wavenumber.

(33) (C) Fingerprint region.

The fingerprint region (below 1,500 cm^{-1}) is a complex region that contains many overlapping peaks due to various bending and skeletal vibrations. It is unique to each molecule and can be used to identify specific compounds by comparing spectra with a database.

(34) (C) Wavenumber is directly proportional to frequency.

Wavenumber represents the spatial incidence or the number of wave cycles that occur over a given distance. However, frequency represents the temporal incidence or the number of wave cycles that occur in a given time. They are directly proportional.

(35) (B) Above 1,500 cm^{-1}.

Functional group region (above 1,500 cm^{-1}) generally has fewer, more distinct peaks that correspond to stretching vibrations of common functional groups. This characteristic makes it more reliable for identifying specific functional groups.

(36) (A) Initiation, propagation, termination.

Radical reactions introduce a new category of reactive intermediates known as radicals, which are species containing an unpaired electron. Radical reactions often follow a chain mechanism, which consists of three main stages: initiation, propagation, and termination.

(37) (C) Homolytic cleavage of bonds.

Homolytic cleavage breaks a covalent bond in such a way that each atom retains one of the shared electrons. This can occur through heat (thermolysis), light (photolysis), or the introduction of a radical initiator. Common initiators include peroxides (e.g., benzoyl peroxide, which decomposes to form radicals), azo compounds (e.g., AIBN), and halogens under UV light.

(38) (B) It removes radicals and halts the reaction.

The termination stage involves the removal of radicals, usually through the combination of two radicals or their reaction with an inhibitor. It reduces the concentration of radicals and eventually halts the chain reaction.

(39) (B) Radicals react with stable molecules to form new radicals.

In the propagation stage, radicals react with stable molecules to generate new radicals, which continue the chain reaction. The overall number of radicals remains constant during these steps, as one radical is consumed when a new radical is generated.

(40) (D) It remains constant.

The overall concentration of radicals remains constant during the propagation stage of radical chain reactions. This is because each radical generated is balanced by a radical that reacts with a stable molecule.

(41) (A) They have unique stability due to delocalized π electrons.

Aromatic compounds are characterized by their unique stability and reactivity. This stability arises from the delocalization of π electrons within a cyclic, planar structure, usually following Hückel's rule ($4n+2$ π electrons).

(42) (C) A base abstracts a proton and restores aromaticity.

In the deprotonation step, a base (usually a conjugate base of the acid catalyst used to generate the electrophile) abstracts a proton from the carbon atom that is now bonded to the electrophile. This deprotonation restores the aromatic π system, which yields the substituted aromatic product and regenerates a proton.

(43) (C) Restoration of aromaticity.

The deprotonation step is usually fast because the driving force is the restoration of aromaticity, which is thermodynamically favorable.

(44) (B) Substitution reactions.

Unlike alkenes, which readily undergo addition reactions, aromatic compounds usually undergo substitution reactions. This preference for substitution preserves the aromaticity of the ring, which maintains its inherent stability.

(45) (B) Electrophilic attack.

The electrophile (E^+) is attracted to the delocalized π electrons of the aromatic ring. It forms a σ bond with one of the carbon atoms. This bond disrupts the aromaticity and creates a positively charged intermediate known as the arenium ion, sigma complex, or Wheland intermediate.

(46) (B) RCHO.

Aldehydes and ketones are two closely related classes of carbonyl compounds that differ in their substitution pattern at the carbonyl carbon. Aldehydes have the general formula RCHO, where R is an alkyl or aryl group, and one substituent is always a hydrogen atom. Ketones have the general formula RCOR', where R and R' are alkyl or aryl groups.

(47) (C) It has a partial positive charge on carbon.

Carbonyl compounds are characterized by the presence of a carbonyl group (C=O). The carbon-oxygen double bond is highly polarized due to the greater electronegativity of oxygen compared to carbon. This polarization results in a partial positive charge (δ+) on the carbon atom and a partial negative charge (δ−) on the oxygen atom.

(48) (A) Solubility decreases.

Lower molecular weight aldehydes and ketones are soluble in water due to their ability to form hydrogen bonds with water molecules. As the size of the alkyl or aryl groups increases, the solubility in water decreases due to the increasing hydrophobic character of the molecule.

(49) (D) Dipole-dipole interactions.

Aldehydes and ketones have higher boiling points than alkanes of comparable molecular weight due to dipole-dipole interactions between the carbonyl groups.

(50) (A) They cannot form hydrogen bonds with each other.

Aldehydes and ketones generally have lower boiling points than alcohols of comparable molecular weight because they cannot form strong hydrogen bonds with themselves.

(51) (B) Structural isomerism.

Keto-enol tautomerism is a type of constitutional isomerism where two isomers, the keto and enol forms, are interconverted through the migration of a proton and the relocation of a double bond. The keto form is a carbonyl compound, while the enol form is an alcohol with a double bond adjacent to the alcohol group (an ene-ol).

(52) (A) Protonation of the carbonyl oxygen.

In acid-catalyzed tautomerism, the process begins with the protonation of the carbonyl oxygen by an acid, which enhances the electrophilicity of the carbonyl carbon. This protonation step is critical as it stabilizes the carbonyl group, which makes it more susceptible to nucleophilic attack.

(53) (C) An alcohol with a double bond adjacent to it.

The enol form is characterized by the presence of a hydroxyl group (−OH) directly attached to a carbon atom that is also involved in a double bond (C=C). This structure contrasts with the keto form, which contains a carbonyl group (C=O).

(54) (A) The keto form is more stable than the enol form in most cases.

In most cases, the keto form is thermodynamically more stable than the enol form and is, therefore, the predominant species at equilibrium. This is due to the greater strength of the C=O double bond compared to the C=C double bond. However, in some instances, the enol form can be stabilized by factors such as conjugation, hydrogen bonding, and aromaticity.

(55) (B) It can shift the equilibrium toward the keto or enol form depending on polarity.

The solvent can influence the equilibrium between the keto and enol forms by affecting their polarity. Polar solvents may stabilize the charged enol form, which potentially favors its formation, while nonpolar solvents may stabilize the keto form.

(56) (A) Loss of electrons or increase in oxidation state.

Oxidation and reduction are often referred to as redox reactions. They are fundamental processes in organic chemistry. They involve the transfer of electrons between reactants, which results in changes in oxidation states. Oxidation is defined as the loss of electrons or an increase in oxidation state, while reduction is defined as the gain of electrons or a decrease in oxidation state. These reactions interconvert functional groups, construct complex molecules, and aid in understanding biological processes.

(57) (B) Formation of a chromate ester.

The mechanism of oxidation with PCC is as follows:

- Formation of chromate ester: The alcohol reacts with PCC to form a chromate ester. This involves the coordination of the alcohol oxygen to the chromium atom.
- Elimination: A proton is abstracted from the carbon atom that bears the oxygen. This eliminates a chromium (IV) species and forms the carbonyl compound (aldehyde or ketone).

(58) (C) It is mild and selective.

PCC is a mild oxidizing agent that does not usually over-oxidize aldehydes to carboxylic acids. This is a significant advantage over more potent oxidizing agents, such as CrO_3 or $KMnO_4$.

(59) (D) It is a stoichiometric reagent.

A limiting factor of PCC is that it is a stoichiometric reagent, which means that one mole of PCC is required for each mole of alcohol oxidized. Another issue is that chromium compounds are toxic and must be handled with care.

(60) (C) Primary alcohols.

Chromium trioxide (CrO_3) is also known as chromic acid. It is a powerful oxidizing agent that can oxidize primary alcohols to carboxylic acids and secondary alcohols to ketones. It is usually used in aqueous sulfuric acid (H_2SO_4) or acetic acid (CH_3COOH) as a solvent.

(61) (B) The mental breaking of a bond in the target molecule.

Retrosynthetic analysis is a problem-solving technique used to plan organic syntheses. It involves working backward from the target molecule to identify simpler starting materials and reaction sequences that can be used to construct the target. The key idea is to break down the target molecule into smaller, more manageable fragments by mentally disconnecting bonds. Disconnection refers to the mental breaking of a bond in the target molecule.

(62) (D) An idealized fragment of a molecule that results from a disconnection.

Organic synthesis is the art and science of constructing organic molecules from simpler building blocks. It is a central discipline in chemistry, with applications that range from the development of new pharmaceuticals and materials to the understanding of complex biological processes. Multistep synthesis involves the sequential execution of multiple chemical reactions to achieve the desired target molecule. The design of efficient and effective multistep syntheses requires a deep understanding of organic reactions, functional group chemistry, and strategic planning. A synthon refers to an idealized fragment of a molecule that results from a disconnection. Synthons are usually charged species (cations or anions) and are not always stable or readily available.

(63) (A) Analyze the target molecule.

The first step in retrosynthetic analysis is to analyze the target molecule carefully. This involves identifying its functional groups, stereocenters, and any unusual structural features. Understanding the molecular architecture is critical, as these elements dictate the reactivity and the types of reactions that can be employed. By recognizing the overall structure, chemists can establish a foundation for the subsequent steps in the analysis.

(64) (A) Functional groups, stereocenters, and structural features.

Identification of functional groups, stereocenters, and structural features is vital because these elements dictate how the molecule will react and what transformations are possible during synthesis.

(65) (C) These locations are usually sites of reactivity.

After the identification of functional groups, the focus shifts to identifying key bonds within the target molecule that can be disconnected. The goal is to simplify the molecule and facilitate the pathway toward readily available starting materials. Several factors should be taken into consideration during this process. Disconnecting bonds adjacent to functional groups is often a strategic approach, as these locations are usually sites of reactivity. This can result in simpler intermediates that can be further elaborated.

(66) (C) Identify a biological target.

Organic chemistry is the study of carbon-containing compounds. It is not confined to the laboratory. It permeates nearly every aspect of modern life, from the medicines we take to the materials that surround us. The pharmaceutical industry relies heavily on organic chemistry for the discovery, development, and synthesis of new drugs. Organic chemists design and synthesize molecules that interact with specific biological targets, such as enzymes, receptors, and DNA, to treat or prevent diseases.

The first step in drug discovery is to identify a biological target that is relevant to the disease of interest. This often involves studying the molecular mechanisms of the disease and identifying key proteins or genes that are involved in its development. Once a target is identified, it must be validated to ensure that modulating its activity will have the desired therapeutic effect. Once a target is validated, the next step is to identify lead compounds that can interact with the target.

(67) (A) Test small libraries of compounds against a target.

Once a target is validated, the next step is to identify lead compounds that can interact with the target. HTS screens large libraries of compounds against the target to identify those that bind or modulate its activity.

(68) (A) Identifies small fragments that bind to the target and links them.

FBDD identifies small, low-affinity fragments that bind to the target and then links them together to create a higher-affinity lead compound. Structure-based drug design utilizes the three-dimensional structure of the target to design molecules that bind to it with high affinity and selectivity.

(69) (A) Absorption, distribution, metabolism, excretion.

Once a lead compound is identified, it must be optimized to enhance its potency, selectivity, and pharmacokinetic properties, which include absorption, distribution, metabolism, and excretion (ADME). This often involves synthesizing and testing a series of analogs of the lead compound.

(70) (C) Preclinical studies.

Before a drug can be tested in humans, it must undergo preclinical studies in animals to assess its safety and efficacy. These studies also provide information about the drug's pharmacokinetic properties. If the preclinical studies are successful, the drug can be tested in humans in clinical trials. Once a drug has been approved by regulatory agencies (e.g., the U.S. Food and Drug Administration), it can be manufactured on a large scale for commercial use. Organic chemists play a key role in developing efficient and cost-effective synthetic routes for drug manufacturing.

Test 2: Questions

(1) In the context of IUPAC prioritization, which functional group has the highest priority over others?

(A) Ketones, which contain a carbonyl group.

(B) Alcohols that are characterized by a hydroxyl group.

(C) Carboxylic acids, known for their –COOH functional group.

(D) Alkenes, which contain a carbon-carbon double bond.

(2) In the compound 4-hydroxy-3-methylpentanoic acid, which functional group determines the suffix used in its name?

(A) The alcohol group.

(B) The methyl group.

(C) The carboxylic acid group.

(D) The pentane chain that forms the backbone of the compound.

(3) What does the numbering of the parent chain ensure in 4-hydroxy-3-methylpentanoic acid?

(A) The alcohol group receives the lowest possible number.

(B) The carboxylic acid carbon receives the lowest possible locant (position 1).

(C) The locant number for the methyl group is minimized.

(D) The length of the carbon chain is maximized.

(4) Which of the following statements correctly describes a carbon-carbon double bond?

(A) It consists of one sigma bond and one pi bond.

(B) It consists of two sigma bonds.

(C) It is longer and weaker than a single bond.

(D) It allows for free rotation around the bond axis.

(5) What is the proper IUPAC name for an alkane consisting of six carbon atoms with a methyl group attached to the third carbon?

(A) 3-methylhexane.

(B) 4-methylhexane.

(C) 3-methylpentane.

(D) 2-methylhexane.

(6) What best describes the phenomenon of resonance in organic molecules?

(A) The movement of nuclei between different structural forms.

(B) The delocalization of π electrons or lone pairs across the molecule.

(C) The establishment of equilibrium between different molecular forms.

(D) The formation of σ bonds between atoms.

(7) Which of the following molecules exhibits resonance?

(A) The nitrate ion (NO_3^-).

(B) Ethane (C_2H_6).

(C) Methane (CH_4).

(D) Butane (C_4H_{10}).

(8) What aspect must remain fixed when depicting resonance structures?

(A) The positions of electrons across different structures.

(B) The positions of atomic nuclei or the atoms themselves.

(C) The formal charges assigned to atoms in the molecule.

(D) The bond angles between atoms in the structure.

(9) Which resonance structure is generally considered to be the most stable?

(A) One that features maximum formal charges on atoms.

(B) One that places negative charges on less electronegative atoms.

(C) One that minimizes formal charges across the structure.

(D) One that violates the octet rule by having broken octets.

(10) How does resonance influence the reactivity of a molecule?

(A) It increases the reactivity of the molecule significantly.

(B) It reduces the reactivity of the molecule.

(C) It does not affect the reactivity whatsoever.

(D) It only impacts the polarity of the molecule.

(11) According to the Brønsted-Lowry theory, what does an acid become after it donates a proton to a base?

(A) A conjugate base.

(B) A conjugate acid.

(C) A hydronium ion.

(D) A hydroxide ion.

(12) Which of the following results from the reaction between a Lewis acid and a Lewis base?

(A) A hydrogen bond.

(B) A coordinate covalent bond.

(C) A metallic bond.

(D) An ionic bond.

(13) What does the acid dissociation constant (K_a) measure?

(A) Strength of a base.

(B) Concentration of hydroxide ions.

(C) Strength of an acid.

(D) Concentration of water.

(14) What does a larger K_a value indicate for an acid?

(A) Weaker acid.

(B) Lower dissociation in solution.

(C) Greater dissociation in solution.

(D) Lower concentration of H_3O^+.

(15) What is the relationship between K_a and pK_a for acids?

(A) Inverse: stronger acids have smaller pK_a values.

(B) Direct: stronger acids have larger pK_a values.

(C) Inverse: weaker acids have smaller pK_a values.

(D) Direct: weaker acids have smaller pK_a values.

(16) How does the presence of different functional groups in functional isomers impact their reactivity?

(A) It does not affect reactivity.

(B) It causes similar reactivity profiles.

(C) It results in significantly different chemical reactivities.

(D) It influences only physical properties.

(17) Which of the following statements regarding the main distinction between constitutional isomers and stereoisomers is correct?

(A) Constitutional isomers have different atoms, while stereoisomers have the same atoms.

(B) Constitutional isomers differ in connectivity, while stereoisomers differ in their spatial arrangement.

(C) Constitutional isomers have different molecular formulas, while stereoisomers have the same molecular formula.

(D) Constitutional isomers have different functional groups, while stereoisomers have the same functional groups.

(18) What impact can the three-dimensional arrangement differences in stereoisomers have on their chemical and physical properties?

(A) No impact.

(B) Identical properties.

(C) Significant differences.

(D) Only minor variations.

(19) Which of the following are a specific type of stereoisomer that are non-superimposable mirror images of each other?

(A) Tristereomers.

(B) Enantiomers.

(C) Diastereomers.

(D) Isostereomers.

(20) Which of the following characteristics is essential for a molecule to exhibit chirality?

(A) A plane of symmetry.

(B) A center of symmetry.

(C) An axis of symmetry.

(D) Lack of symmetrical elements.

(21) What happens to the configuration at the stereocenter in an SN2 reaction?

(A) Retains the configuration.

(B) No change in the configuration.

(C) Inverts the configuration.

(D) Forms a racemic mixture.

(22) What is the first step of the SN1 reaction mechanism?

(A) Nucleophilic attack.

(B) Formation of a carbocation.

(C) Departure of the leaving group.

(D) Solvation of the carbocation.

(23) Which type of substrate is most likely to undergo an SN1 reaction?

(A) Primary substrates.

(B) Secondary substrates.

(C) Tertiary substrates.

(D) Quaternary substrates.

(24) What is the rate law for an SN1 reaction?

(A) Rate = k[substrate].

(B) Rate = k[substrate]2.

(C) Rate = k[substrate] [nucleophile]2.

(D) Rate = k[substrate][nucleophile].

(25) What stereochemical outcome is associated with the SN1 mechanism?

(A) Retention of configuration.

(B) Inversion of configuration.

(C) No stereochemical change.

(D) Racemization.

(26) Which of the following is regarded as the first step in the acid-catalyzed hydration of an alkene?

(A) Nucleophilic attack.

(B) Deprotonation.

(C) Protonation.

(D) Formation of the product.

(27) What is the final product after deprotonation of the protonated alcohol during the acid-catalyzed hydration of an alkene?

(A) Alkene.

(B) Neutral alcohol.

(C) Carbocation.

(D) Water.

(28) What is formed after the nucleophilic attack by water on the carbocation during the acid-catalyzed hydration of an alkene?

(A) Alkene.

(B) Protonated alcohol.

(C) Ether.

(D) Carbocation.

(29) Which of the following reactions involves the addition of water (H_2O) to alkenes to create alcohols?

(A) Hydration.

(B) Protonation.

(C) Deprotonation.

(D) Hydrohalogenation.

(30) Which of the following reactions involves the addition of hydrogen halides (HX, where X = Cl, Br, I) to alkenes?

(A) Hydration.

(B) Protonation.

(C) Deprotonation.

(D) Hydrohalogenation.

(31) Which of the following best describes the structure of a ketone functional group?

(A) A carbon atom double-bonded to an oxygen and single-bonded to a hydrogen.

(B) A carbon atom double-bonded to an oxygen and single-bonded to two other carbon atoms.

(C) A carbon atom bonded to a hydroxyl group and a hydrogen.

(D) A carbon atom double-bonded to two oxygen atoms.

(32) Which functional group exhibits a broad O–H stretch in the range of 3200–3650 cm^{-1}?

(A) Alcohols and phenols.

(B) Carboxylic Acids.

(C) Aldehydes.

(D) Alkenes.

(33) Which of the following statements regarding the effect of bond strength on IR vibrational frequencies is correct?

(A) Stronger bonds vibrate at lower frequencies.

(B) Stronger bonds vibrate at higher frequencies.

(C) Bond strength does not affect frequency.

(D) All bonds vibrate at the same frequency.

(34) Which hybridization state corresponds to the lowest C–H stretching frequency?

(A) sp.

(B) sp^2.

(C) sp^3.

(D) sp^4.

(35) Which of the following statements regarding the relationship between atomic mass and vibrational frequency is correct?

(A) Lighter atoms vibrate at lower frequencies.

(B) Heavier atoms vibrate at higher frequencies.

(C) Lighter atoms vibrate at higher frequencies.

(D) Atomic mass does not affect vibrational frequency.

(36) What type of mechanism do alkanes undergo when reacting with halogens?

(A) Ionic.

(B) Addition.

(C) Elimination.

(D) Radical chain.

(37) What does a chlorine radical do during the propagation step in the halogenation of alkanes?

(A) Form a stable compound.

(B) Abstract a hydrogen atom from the alkane.

(C) React with another chlorine radical.

(D) Initiate decomposition of alkanes.

(38) What is the correct order of reactivity for hydrogen atoms in alkanes during halogenation?

(A) Primary > secondary > tertiary.

(B) Secondary > primary > tertiary.

(C) Tertiary > secondary > primary.

(D) All display equally reactive.

(39) Why is bromination of alkanes more selective than chlorination?

(A) Bromine is more abundant.

(B) Bromine radicals are less stable.

(C) Greater stability of the bromine radical and transition state.

(D) Chlorination requires higher temperatures.

(40) What happens during the initiation step in the halogenation of alkanes?

(A) Homolytic cleavage of Cl_2 by UV light.

(B) Abstraction of a hydrogen atom from the alkane.

(C) A chlorine radical reacts with another radical.

(D) Decomposition of alkanes.

(41) All of the following are names of the positively charged intermediate formed during the electrophilic attack stage of EAS except:

(A) Carbocation.

(B) Sigma complex.

(C) Arenium ion.

(D) Wheland intermediate.

(42) Which of the following acids are usually used for the nitration of aromatic compounds?

(A) Hydrochloric and acetic acid.

(B) Phosphoric and citric acid.

(C) Sulfuric and hydrochloric acid.

(D) Nitric and sulfuric acid.

(43) What role does sulfuric acid play in the nitration process of aromatic rings?

(A) It protonates nitric acid to form the nitronium ion.

(B) It neutralizes any byproducts.

(C) It directly attacks the aromatic ring.

(D) It acts as a reducing agent.

(44) What occurs during the deprotonation step of the nitration of aromatic rings?

(A) A nitro group is lost.

(B) Aromaticity is restored.

(C) Water is produced.

(D) A new electrophile is generated.

(45) What functional group is introduced during the sulfonation of an aromatic ring?

(A) Hydroxyl group (–OH).

(B) Carbonyl group (C=O).

(C) Sulfonic acid group ($-SO_3H$).

(D) Sulfur trioxide (SO_3).

(46) Which of the following aldehydes is known for its pleasant odor?

(A) Formaldehyde.

(B) Benzaldehyde.

(C) Acetaldehyde.

(D) Propionaldehyde.

(47) What is the characteristic absorption band for the carbonyl group in IR spectroscopy?

(A) 1600–1650 cm^{-1}.

(B) 1700–1750 cm^{-1}.

(C) 2800–3000 cm^{-1}.

(D) 2000–2100 cm^{-1}.

(48) What type of reaction is characterized by nucleophiles attacking the carbonyl carbon of a carbonyl group?

(A) Nucleophilic addition.

(B) Electrophilic substitution.

(C) Elimination reaction.

(D) Rearrangement reaction.

(49) What is the first step in the nucleophilic addition mechanism to a carbonyl group?

(A) Protonate the tetrahedral intermediate.

(B) Form a π bond.

(C) Break the σ bond.

(D) A nucleophilic attack on the carbonyl carbon.

(50) What product is formed when water adds to a carbonyl group during the process of hydration?

(A) Aldehyde.

(B) Ketone.

(C) Gem-diol.

(D) Acetal.

(51) What mainly contributes to the greater stability of the keto form over the enol form?

(A) Conjugation.

(B) Hydrogen bonding.

(C) Strength of the C=O double bond.

(D) Aromaticity.

(52) What happens to the equilibrium position when the enol form is stabilized by hydrogen bonding?

(A) It shifts toward the keto form.

(B) It remains unchanged.

(C) It shifts toward the enol form.

(D) It becomes impossible to determine.

(53) All of the following factors contribute to the stability of the enol form except:

(A) Conjugation.

(B) Hydrogen bonding.

(C) Aromaticity.

(D) Increased steric strain.

(54) What advantage do bulky and non-nucleophilic bases have in enolate formation?

(A) They can deprotonate multiple sites simultaneously.

(B) They can selectively deprotonate the less hindered α-carbon.

(C) They are more reactive than smaller bases.

(D) They do not affect the regioselectivity of the reaction.

(55) Under what conditions is a kinetic enolate usually formed?

(A) Low temperatures with a strong, bulky base.

(B) High temperatures with a weak base.

(C) Room temperature with a strong acid.

(D) High temperatures with a strong nucleophile.

(56) What type of alcohols does CrO_3 oxidize to ketones?

(A) Tertiary alcohols.

(B) Secondary alcohols.

(C) Primary alcohols.

(D) All types of alcohols.

(57) Which statement correctly describes the Jones reagent?

(A) A solution of CrO_3 in dichloromethane.

(B) A solution of CrO_3 in aqueous sulfuric acid.

(C) A complex of CrO_3 with pyridine.

(D) A solution of CrO_3 in ethanol.

(58) What is a significant limitation of using CrO_3 as an oxidizing agent?

(A) It is ineffective at oxidizing alcohols.

(B) It is always safe to handle.

(C) It requires low temperatures for reactions.

(D) It can over-oxidize aldehydes to carboxylic acids.

(59) What is the first step in the mechanism of oxidation with CrO_3?

(A) Formation of a chromate ester.

(B) Deprotonation of the alcohol.

(C) Reduction to a carbonyl compound.

(D) Elimination of water.

(60) What type of agent is potassium permanganate ($KMnO_4$)?

(A) Reducing agent.

(B) Neutral agent.

(C) Strong oxidizing agent.

(D) Weak oxidizing agent.

(61) How should the disconnection process be handled during retrosynthetic analysis?

(A) It should be done only once.

(B) It should be repeated iteratively until starting materials are identified.

(C) It should only focus on one bond at a time.

(D) It should avoid considering symmetry.

(62) Which statement regarding the final step in retrosynthetic analysis is correct?

(A) Reanalyze the target molecule.

(B) Write the forward synthesis.

(C) Propose additional synthons.

(D) Identify more reagents.

(63) Why is it important to start with the most complex part of the molecule in retrosynthetic analysis?

(A) It simplifies the entire synthesis process.

(B) It reduces the number of reactions needed.

(C) It eliminates the need for protecting groups.

(D) It makes the synthesis less challenging.

(64) What action do functional group interconversions (FGIs) perform in retrosynthetic analysis?

(A) Transform functional groups into more useful or reactive forms.

(B) Isolate pure compounds.

(C) Combine different molecules.

(D) Analyze the stability of compounds.

(65) Why are protecting groups important in retrosynthetic analysis?

(A) They temporarily mask reactive functional groups to prevent unwanted reactions.

(B) They speed up the reaction rate.

(C) They simplify the final product.

(D) They enhance the color of the product.

(66) What phase involves large, randomized, controlled trials to confirm a drug's efficacy and monitor long-term safety?

(A) Phase I.

(B) Phase II.

(C) Phase III.

(D) Phase IV.

(67) Which of the following antibiotics was discovered by Alexander Fleming?

(A) Aspirin.

(B) Taxol.

(C) Penicillin.

(D) Ibuprofen.

(68) Which of the following statements regarding the constituents of polymers is correct?

(A) Single atoms.

(B) Random chemical compounds.

(C) Metal ions.

(D) Repeating structural units called monomers.

(69) Which of the following statements correctly characterizes addition polymerization?

(A) Monomers join together with the loss of a small molecule.

(B) Monomers add to each other without the loss of atoms.

(C) Monomers join together due to protein formation.

(D) It involves diacids and diamines.

(70) Which of the following statements correctly describes the glass transition temperature (Tg) of a polymer?

(A) The temperature at which a polymer melts.

(B) The temperature at which a polymer becomes brittle.

(C) The temperature at which a polymer transitions from a rigid to a rubbery state.

(D) The temperature at which a polymer decomposes.

Test 2: Answers and Explanations

(1) (C) Carboxylic acids, known for their –COOH functional group.

Carboxylic acids are assigned the highest priority in IUPAC nomenclature, which ranks them above other functional groups such as ketones, alcohols, and alkenes. The prioritization in IUPAC nomenclature plays a vital role in how organic compounds are named and categorized. This protocol ensures clarity and consistency in chemical communication.

(2) (C) The carboxylic acid group.

The suffix -oic acid is determined by the presence of the carboxylic acid, which is the highest-priority functional group in this compound. The alcohol group is indicated as a lower-priority prefix, while the methyl group is simply a substituent and does not dictate the suffix.

(3) (B) The carboxylic acid carbon receives the lowest possible locant (position 1).

The numbering of the parent chain is performed in such a way that the highest-priority functional group, which is the carboxylic acid, receives the lowest possible locant, specifically position 1. The alcohol group is considered lower in priority. Although it is important to minimize locants for substituents, it is secondary to prioritizing the carboxylic acid.

(4) (A) It consists of one sigma bond and one pi bond.

A carbon-carbon double bond is made up of one sigma (σ) bond, formed by head-on overlap of orbitals, and one pi (π) bond, formed by the side-by-side overlap of p orbitals. The pi bond restricts rotation, giving the double bond rigid geometry.

(5) (A) 3-methylhexane.

The longest carbon chain in this compound contains six carbon atoms, and the methyl group is correctly located on the third carbon. This order results in the name 3-methylhexane. The other options misplace the methyl group or misidentify the length of the parent chain.

(6) (B) The delocalization of π electrons or lone pairs across the molecule.

Resonance in organic chemistry refers specifically to the delocalization of π electrons or lone pairs across conjugated systems, which serves to stabilize the molecule. The movement of nuclei between different structural forms is inaccurate since the positions of atomic nuclei remain fixed in resonance structures. The establishment of equilibrium between different molecular forms confuses resonance with dynamic equilibrium, and the formation of σ bonds between atoms focuses on bonding rather than the concept of resonance itself.

(7) (A) The nitrate ion (NO_3^-).

The nitrate ion has three equivalent resonance structures that feature delocalized double bonds among oxygen atoms. Ethane (C_2H_6), methane (CH_4), and butane (C_4H_{10}) represent alkanes, which contain only single σ bonds and do not exhibit resonance.

(8) (B) The positions of atomic nuclei or the atoms themselves.

In resonance structures, the only variations occur in the placement of electrons. The nuclear positions, or the atoms themselves, remain unchanged. The statement that the positions of electrons across different structures is incorrect because electrons can move between resonance forms. The formal charges assigned to atoms in the molecule can vary between resonance structures, and the bond angles between atoms in the structure are a consequence of electronic arrangement, not a fixed rule.

(9) (C) One that minimizes formal charges across the structure.

Minimal formal charges characterize the most stable resonance structure and usually place negative charges on more electronegative atoms, which enhances stability.

(10) (B) It reduces the reactivity of the molecule.

Resonance helps delocalize electrons throughout the molecule. This stabilizes it and consequently reduces its reactivity, particularly in compounds such as aromatic systems. Resonance usually stabilizes rather than increases reactivity.

(11) (A) A conjugate base.

The Brønsted-Lowry acid-base theory emphasizes the relationship between acids and bases. It states that when an acid donates a proton to a base, it forms a conjugate base, and the base becomes a conjugate acid upon accepting the proton.

(12) (B) A coordinate covalent bond.

The reaction between a Lewis acid and a Lewis base results in the formation of a coordinate covalent bond, where both electrons in the bond originate from the Lewis base. The resulting species is known as an adduct or complex. For example, when ammonia (NH_3) reacts with boron trifluoride (BF_3), the nitrogen in ammonia donates its lone pair to the electron-deficient boron atom.

- $NH_3 + BF_3 \longrightarrow H_3N{-}BF_3$: In this reaction, BF_3 acts as a Lewis acid, while NH_3 serves as a Lewis base.

(13) (C) Strength of an acid.

The acid dissociation constant (K_a) quantitatively measures the strength of an acid in solution. This calculation indicates how well an acid (HA) dissociates into its conjugate base (A^-) and a proton (H^+) in water.

(14) (C) Greater dissociation in solution.

A larger K_a value signifies a stronger acid, which indicates greater dissociation in solution and a higher concentration of H_3O^+. For convenience, acidity is often expressed using the pK_a value, which is the negative logarithm (base 10) of K_a.

(15) (A) Inverse: stronger acids have smaller pK_a values.

The relationship between K_a and pK_a is inverse. Stronger acids have higher K_a values and lower pK_a values. Weaker acids have lower K_a values and higher pK_a values.

(16) (C) It results in significantly different chemical reactivities.

Functional isomers usually exhibit significantly different chemical reactivities because the functional group dictates how the molecule interacts with other substances.

For example:

- Ethanol (C_2H_5OH): This compound is an alcohol with a hydroxyl functional group.
- Dimethyl Ether (CH_3OCH_3): This compound is an ether characterized by an oxygen atom bonded to two carbon groups.

The presence of different functional groups not only affects the chemical reactivity but also the physical properties, such as boiling points and solubility of these compounds.

(17) (B) Constitutional isomers differ in connectivity, while stereoisomers differ in their spatial arrangement.

Constitutional isomers have the same molecular formula but differ in how their atoms are connected. This variation in connectivity can result in different functional groups, carbon skeletons, or positions of substituents. Stereoisomers are molecules that have the same molecular formula (number and types of atoms) and the same connectivity (sequence of bonded atoms) but differ in the three-dimensional arrangement of their atoms in space.

(18) (C) Significant differences.

The spatial difference in stereoisomers results in distinct chemical and physical properties, despite their identical composition. Stereoisomers are important in fields such as organic chemistry, biochemistry, and pharmacology, as their spatial arrangements can significantly influence their behavior, which includes how they interact with biological systems.

(19) (B) Enantiomers.

Stereoisomers are broadly classified into two main categories based on their structural relationships: enantiomers and diastereomers. Enantiomers are a specific type of stereoisomer that are non-superimposable mirror images of each other, much like a left hand and a right hand.

(20) (D) Lack of symmetrical elements.

Chiral molecules lack symmetrical elements such as a plane, center, or axis of symmetry. Chirality is often associated with a chiral center (or stereocenter), usually a carbon atom bonded to four different substituents.

(21) (C) Inverts the configuration.

The SN2 (substitution nucleophilic bimolecular) reaction involves a single concerted step where the nucleophile attacks the electrophilic carbon atom from the opposite side of the leaving group. This backside attack is vital for the mechanism. When the nucleophile approaches the carbon atom, it repels the leaving group, which inverts the spatial arrangement of the groups attached to that carbon atom.

(22) (C) Departure of the leaving group.

The SN1 reaction occurs in two steps. The first step is the slow, rate-determining departure of the leaving group, which forms a carbocation intermediate. The second step involves a rapid nucleophilic attack on the carbocation.

(23) (C) Tertiary substrates.

The SN1 reaction usually occurs in tertiary substrates due to carbocation stability. The reaction rate depends only on the concentration of the substrate. The formation of the carbocation can cause racemization if the nucleophile can attack from either side.

(24) (A) Rate = k[substrate].

The reaction rate of an SN1 reaction is first order as it depends only on the substrate concentration: Rate = k[substrate].

(25) (D) Racemization.

SN1 mechanism causes racemization at the stereocenter, as the nucleophile can attack the planar carbocation from either face. This process results in a mixture of enantiomers or a slight preference for inversion if the leaving group blocks one side.

(26) (C) Protonation.

Hydration refers to the addition of water (H_2O) to alkenes to create alcohols, usually facilitated by an acid catalyst. Protonation is the first step in which an alkene is protonated by an acid catalyst (e.g., H_3O^+), which forms a carbocation on the more substituted carbon (as per Markovnikov's rule).

(27) (B) Neutral alcohol.

During the last stage (deprotonation), a water molecule removes a proton from the protonated alcohol, which yields the neutral alcohol and regenerates the acid catalyst (H_3O^+).

(28) (B) Protonated alcohol.

During the second stage of the acid-catalyzed hydration of an alkene, a nucleophilic attack happens where water acts as the nucleophile. It attacks the carbocation to form a protonated alcohol ($R{-}OH_2^+$).

(29) (A) Hydration.

Hydration is the process of adding water (H_2O) to alkenes, which results in the formation of alcohols. This reaction usually requires an acid catalyst (like H_3O^+) to protonate the alkene and create a carbocation. Water then acts as a nucleophile and attacks the carbocation to form a protonated alcohol. Finally, deprotonation occurs to yield the neutral alcohol.

(30) (D) Hydrohalogenation.

Hydrohalogenation refers to the addition of hydrogen halides (HX) to alkenes and results in the formation of alkyl halides. In this reaction, the hydrogen from the halide adds to one carbon of the alkene, while the halide (X) adds to the other carbon. Similar to hydration, this process also follows Markovnikov's rule, where the hydrogen atom attaches to the less substituted carbon.

(31) (B) A carbon atom double-bonded to an oxygen and single-bonded to two other carbon atoms.

A ketone has the general structure RCOR′, where the central carbon is double-bonded to an oxygen atom (a carbonyl group) and single-bonded to two alkyl or aryl groups. This distinguishes ketones from aldehydes, which have one carbon and one hydrogen attached to the carbonyl carbon.

(32) (A) Alcohols and phenols.

Alcohols and phenols display a broad O–H stretch in the range of 3200–3650 cm^{-1}. The broadening is due to hydrogen bonding, which affects the vibrational frequency of the O–H bond.

(33) (B) Stronger bonds vibrate at higher frequencies.

The vibrational frequency of a bond is influenced by its strength. Stronger bonds, such as triple bonds (e.g., C≡C), vibrate at higher frequencies compared to weaker bonds (e.g., single bonds like C–C). This is due to the increased energy required to stretch or compress a stronger bond.

(34) (C) sp^3.

The hybridization state determines the C–H stretching frequency. In general, sp-hybridized carbon (which involves a triple bond) has the highest C–H stretching frequency, followed by sp^2 (double bond), and then sp^3 (single bond), which has the lowest frequency.

(35) (C) Lighter atoms vibrate at higher frequencies.

The vibrational frequency of bonds is inversely related to the mass of the atoms involved. Lighter atoms (such as hydrogen) vibrate at higher frequencies compared to heavier atoms (like deuterium or oxygen)

(36) (D) Radical chain.

The reaction of alkanes with halogens (Cl_2 or Br_2) in the presence of light or heat proceeds through a radical chain mechanism. This process results in the substitution of one or more hydrogen atoms with halogen atoms.

(37) (B) Abstract a hydrogen atom from the alkane.

A chlorine radical reacts with the alkane and abstracts a hydrogen atom to form an alkyl radical and a new chlorine radical during the propagation step in the halogenation of alkanes. This will be followed by the termination step of the chain reaction.

(38) (C) Tertiary > secondary > primary.

Halogenation of alkanes is generally not highly regioselective and often yields a mixture of products when alkanes with different hydrogen types (primary, secondary, tertiary) react. However, the reactivity order of hydrogen atoms is tertiary > secondary > primary, which reflects the stability of the resulting alkyl radical.

(39) (C) Greater stability of the bromine radical and transition state.

Bromination is more selective than chlorination due to the greater stability of the bromine radical and the transition state for hydrogen abstraction.

(40) (A) Homolytic cleavage of Cl_2 by UV light.

The initiation step involves the homolytic cleavage of Cl_2 molecules under UV light. This generates two chlorine radicals, which begin the radical chain mechanism for halogenation.

(41) (A) Carbocation.

The electrophile (E^+) is attracted to the delocalized π electrons of the aromatic ring. It forms a σ bond with one of the carbon atoms, which disrupts the aromaticity and creates a positively charged intermediate known as the arenium ion, sigma complex, or Wheland intermediate.

(42) (D) Nitric and sulfuric acid.

Nitration introduces a nitro group ($-NO_2$) onto the aromatic ring. This is usually achieved using a mixture of concentrated nitric acid (HNO_3) and sulfuric acid (H_2SO_4). Sulfuric acid acts as a catalyst by protonating nitric acid, which forms the nitronium ion (NO_2^+), the active electrophile.

(43) (A) It protonates nitric acid to form the nitronium ion.

Sulfuric acid protonates nitric acid, which forms a protonated nitric acid species. This species then loses water to generate the nitronium ion (NO_2^+), the active electrophile.

(44) (B) Aromaticity is restored.

During deprotonation, a base (usually HSO_4^-) removes a proton from the arenium ion, which restores aromaticity and yields nitrobenzene. Nitrobenzene is a key intermediate in the synthesis of various aromatic compounds, particularly amines, through reduction (e.g., using Sn/HCl or H_2/Pd).

(45) (C) Sulfonic acid group (–SO_3H).

Sulfonation introduces a sulfonic acid group (–SO_3H) onto the aromatic ring. This is usually carried out by heating benzene with fuming sulfuric acid (H_2SO_4 containing dissolved SO_3) or concentrated sulfuric acid at elevated temperatures. The electrophile is sulfur trioxide (SO_3).

(46) (B) Benzaldehyde.

Many aldehydes and ketones have characteristic odors. Some are pleasant (e.g., vanillin in vanilla extract, benzaldehyde in almonds), while others are pungent or unpleasant (e.g., formaldehyde).

(47) (B) 1700–1750 cm^{-1}.

The carbonyl group exhibits a strong, characteristic absorption band in the IR spectrum in the region of 1700–1750 cm^{-1}. The exact position of the band depends on the structure of the aldehyde or ketone and the presence of other functional groups. Aldehydes also show characteristic C–H stretching vibrations around 2700–2850 cm^{-1}.

(48) (A) Nucleophilic addition.

The carbonyl group is highly susceptible to nucleophilic attack due to the electrophilic nature of the carbonyl carbon. Nucleophilic addition reactions are a hallmark of carbonyl chemistry.

(49) (D) A nucleophilic attack on the carbonyl carbon.

During a nucleophilic attack, the nucleophile (Nu^-) attacks the electrophilic carbonyl carbon, which forms a new σ bond and breaks the π bond. This generates a tetrahedral intermediate with a negative charge on the oxygen atom.

(50) (C) Gem-diol.

Water can act as a nucleophile and add to the carbonyl group to form a gem-diol (a diol with both hydroxyl groups on the same carbon). This reaction is usually catalyzed by acid or base.

(51) (C) Strength of the C=O double bond.

In most cases, the keto form is thermodynamically more stable than the enol form due to the greater strength of the C=O double bond compared to the C=C double bond. However, in some instances, the enol form can be stabilized by factors such as conjugation, hydrogen bonding, and aromaticity.

(52) (C) It shifts toward the enol form.

Intramolecular hydrogen bonding is an important stabilizing factor, which is particularly prominent in β-dicarbonyl compounds. In these molecules, the enol form can create a stable six-membered ring through hydrogen bonding between the hydroxyl group of the enol and a carbonyl oxygen. This intramolecular interaction not only stabilizes the enol form but can also influence the equilibrium position, which makes it more likely to exist in a significant concentration.

(53) (D) Increased steric strain.

Aromaticity provides a substantial driving force for enolization because the resulting aromatic compound is more stable than its non-aromatic counterparts. In contrast, increased steric strain would make the enol form less stable, as bulky groups would repel each other.

(54) (B) They can selectively deprotonate the less hindered α-carbon.

To achieve greater control over enolate formation, chemists often utilize bulky, non-nucleophilic bases such as lithium diisopropylamide (LDA) and potassium tert-butoxide (KOtBu). These bases are advantageous because their size and steric hindrance allow them to deprotonate the less hindered α-carbon selectively.

(55) (A) Low temperatures with a strong, bulky base.

The kinetic enolate is formed under conditions that favor rapid, irreversible deprotonation. This often involves the use of a strong, bulky base, such as LDA, at low temperatures.

(56) (B) Secondary alcohols.

Chromium trioxide (CrO_3) (chromic acid) is a powerful oxidizing agent that can oxidize primary alcohols to carboxylic acids and secondary alcohols to ketones. It is usually used in aqueous sulfuric acid (H_2SO_4) or acetic acid (CH_3COOH) as a solvent.

(57) (B) A solution of CrO_3 in aqueous sulfuric acid.

A solution of CrO_3 in aqueous sulfuric acid is known as the Jones reagent. It is a very effective oxidizing agent for converting primary alcohols to carboxylic acids. The Collins reagent is a complex of CrO_3 with pyridine in dichloromethane. It is similar to PCC but more difficult to prepare.

(58) (D) It can over-oxidize aldehydes to carboxylic acids.

The following are the limitations of CrO_3.

- Strong oxidizing agent: CrO_3 is a strong oxidizing agent that can over-oxidize aldehydes to carboxylic acids.
- Harsh conditions: The reaction conditions are often harsh, which can result in unwanted side reactions.
- Toxic: Chromium compounds are toxic and must be handled with care.
- Not suitable for acid-sensitive compounds: The strongly acidic conditions can be problematic for molecules containing acid-sensitive functional groups.

(59) (A) Formation of a chromate ester.

The mechanism of oxidation with CrO_3 is as follows:

1) Formation: The alcohol reacts with CrO_3 to form a chromate ester.
2) Elimination: A proton is abstracted from the carbon atom bearing the oxygen, which eliminates a chromium(IV) species and forms the carbonyl compound (aldehyde or ketone).
3) Further Oxidation (for Primary Alcohols): If a primary alcohol is used, the aldehyde intermediate can be further oxidized to a carboxylic acid. This involves the addition of water to the aldehyde, followed by oxidation of the resulting gem-diol.

This process is similar to that of PCC. However, because CrO_3 is a stronger oxidizing agent, the reaction usually proceeds further than with PCC.

(60) (C) Strong oxidizing agent.

Potassium permanganate ($KMnO_4$) is a strong oxidizing agent that can oxidize a wide variety of organic compounds. It is usually used in aqueous solution, either under acidic or basic conditions.

(61) (B) It should be repeated iteratively until starting materials are identified.

During retrosynthetic analysis, the disconnection process should be repeated iteratively until it yields readily available starting materials. Each cycle of disconnection and identification of synthons and reagents should progressively simplify the target molecule. This process ultimately guides chemists to the most practical synthetic pathway.

(62) (B) Write the forward synthesis.

Once a comprehensive retrosynthetic analysis has been completed, the last step is to write out the forward synthesis. The chemist details the synthetic route, which includes the reagents and conditions for each step. The forward synthesis serves as a roadmap for the experimental work required to construct the target molecule from its starting materials.

(63) (A) It simplifies the entire synthesis process.

A fundamental guideline in retrosynthetic analysis is to begin with the most complex part of the target molecule. Chemists focus on the most challenging structural features first, which can simplify the synthesis process. This approach enables the identification of critical bonds that, when disconnected, can yield more manageable intermediates. Tackling complexity early in the analysis often facilitates the overall synthetic strategy, as it reduces the likelihood of encountering unforeseen difficulties in the future.

(64) (A) Transform functional groups into more useful or reactive forms.

FGIs are essential tools in retrosynthetic analysis. These conversions allow chemists to transform functional groups into more useful or reactive forms, which enhances the molecule's synthetic accessibility. By strategically planning FGIs, chemists can create intermediates that are better suited for subsequent transformations. This flexibility helps navigate the complexities of organic synthesis, where the reactivity of functional groups can dictate the success of a reaction.

(65) (A) They temporarily mask reactive functional groups to prevent unwanted reactions.

In many synthetic pathways, certain functional groups may react in ways that interfere with desired transformations. Protecting groups can temporarily mask reactive functional groups, which allows chemists to carry out specific reactions without unwanted side effects. By incorporating protecting groups into the retrosynthetic analysis, chemists can maintain control over the reaction conditions and ensure that the desired transformations proceed smoothly.

(66) (C) Phase III.

Clinical trials are usually conducted in three phases.

- Phase I trials involve small studies conducted with healthy volunteers to assess a drug's safety and pharmacokinetic properties.
- Phase II trials are larger studies that involve patients with the disease of interest. They evaluate the drug's efficacy and side effects.
- Phase III trials consist of large, randomized, controlled studies. They aim to confirm the drug's efficacy and monitor its long-term safety before it can be approved for general use.

Once a drug has been approved by regulatory agencies (e.g., the U.S. Food and Drug Administration), it can be manufactured on a large scale for commercial use. Organic chemists work to develop efficient and cost-effective synthetic routes for drug manufacturing.

(67) (C) Penicillin.

Penicillin is a natural product antibiotic discovered by Alexander Fleming. Organic chemists have developed methods for synthesizing penicillin and its analogs on a large scale.

(68) (D) Repeating structural units called monomers.

Polymers are large molecules made up of repeating structural units called monomers. They are classified as either natural polymers (e.g., cellulose, starch, proteins) or synthetic polymers (e.g., polyethylene, polystyrene, nylon).

(69) (B) Monomers add to each other without the loss of atoms.

There are two main types of polymerization.

- Addition polymerization: Monomers add to each other without the loss of any atoms. Examples include the polymerization of ethylene to form polyethylene and the polymerization of styrene to form polystyrene.
- Condensation polymerization: Monomers join together with the loss of a small molecule, such as water or methanol. Examples include the polymerization of amino acids to form proteins and the polymerization of diacids and diamines to form polyamides (nylons).

(70) (C) The temperature at which a polymer transitions from a rigid to a rubbery state.

The properties of a polymer depend on its chemical structure, molecular weight, and morphology (the arrangement of the polymer chains). Some important properties of polymers include the following.

- Tensile strength: The ability of a polymer to resist stretching or breaking.
- Elasticity: The ability of a polymer to return to its original shape after being deformed.
- Glass transition temperature (Tg): The temperature at which a polymer transitions from a rigid, glassy state to a rubbery state.
- Crystallinity: The degree to which the polymer chains are ordered in a crystalline structure.

Test 3: Questions

(1) What is the general molecular formula that applies to cycloalkanes that contain a single-ring structure?

(A) C_nH_{2n+2}.

(B) C_nH_{2n}.

(C) C_nH_{2n-2}.

(D) C_nH_n.

(2) What accounts for the decreased stability of cyclopropane and cyclobutane compared to cyclohexane?

(A) They exhibit higher boiling points than cyclohexane.

(B) They are subject to angle strain due to their bond angles.

(C) They possess fewer hydrogen atoms than cyclohexane.

(D) They have a higher degree of saturation than cyclohexane.

(3) Which conformation of cyclohexane is recognized as being the most stable?

(A) The boat conformation.

(B) The chair conformation.

(C) The twist-boat conformation.

(D) The half-chair conformation.

(4) Why do alkynes have higher boiling points than alkenes of comparable molecular weight despite both having lower boiling points than their saturated counterparts?

(A) Alkynes have weaker London dispersion forces due to the triple bond.

(B) The triple bond in alkynes increases the strength of intermolecular forces.

(C) Alkenes have stronger dipole-dipole interactions than alkynes.

(D) Alkynes are saturated and cause stronger Van der Waals forces.

(5) What is the correct IUPAC name for an alkene with six carbon atoms featuring a double bond between carbons 2 and 3?

(A) Hex-3-ene.

(B) Hex-2-ene.

(C) Hex-1-ene.

(D) Hex-4-ene.

(6) How does resonance affect the acidity of phenolic compounds?

(A) It destabilizes the resulting conjugate base after deprotonation.

(B) It stabilizes the conjugate base and enhances acidity.

(C) It increases the retention of protons within the molecule.

(D) It decreases the overall electron density in the structure.

(7) What is a common mistake made when drawing resonance structures?

(A) Include all lone pairs of electrons in the structures.

(B) Move sigma (σ) bonded atoms or alter σ bonds.

(C) Adhere to the octet rule for all atoms involved.

(D) Accurately calculate the formal charges on atoms.

(8) What is a necessary condition for a molecule to be classified as aromatic?

(A) The molecule must be acyclic, which means it cannot be a ring.

(B) The molecule must possess a planar structure.

(C) The molecule must contain 4n π electrons.

(D) The molecule must lack any form of conjugation.

(9) How many π electrons are present in furan, which is recognized as an aromatic compound?

(A) Four.

(B) Six.

(C) Eight.

(D) Ten.

(10) Why is cyclopentadiene considered to be non-aromatic?

(A) It does not have a cyclic structure.

(B) It lacks full conjugation due to structural constraints.

(C) It is planar, which usually supports aromaticity.

(D) It possesses six π electrons, which is characteristic of aromatic compounds.

(11) What relationship exists between pK_a and pK_β for a conjugate acid-base pair in aqueous solution?

(A) $pK_a + pK_\beta = pK_h = 14$.

(B) $pK_a - pK_\beta = pK_h = 7$.

(C) $pK_a + pK_\beta = pK_h = 7$.

(D) $pK_a - pK_\beta = pK_h = 14$.

(12) Which of the following statements best explains why a weaker acid corresponds to a stronger base?

(A) Weaker acids dissociate more completely in solution.

(B) The conjugate base of a weaker acid is more stable.

(C) The conjugate base of a weaker acid is less stable and more willing to accept protons.

(D) Weaker acids have lower pKa values.

(13) What does a lower pK_a value indicate for an acid?

(A) Weaker acid.

(B) Stronger acid.

(C) Lower dissociation in solution.

(D) Lower concentration of H_3O^+.

(14) Which of the following statements regarding the relationship between acidity and electronegativity within the same period of the periodic table is correct?

(A) Acidity increases as electronegativity increases.

(B) Acidity decreases as electronegativity increases.

(C) Acidity remains constant.

(D) Acidity is not affected by electronegativity.

(15) What specific property allows enantiomers to be referred to as optical isomers?

(A) Identical physical properties.

(B) Mirror-image relationship.

(C) Interaction with chiral environments.

(D) Rotation of plane-polarized light in opposite directions.

(16) Which of the following is the term for a 50:50 mixture of enantiomers where the rotations of plane-polarized light cancel out?

(A) Racemic mixture.

(B) Chiral mixture.

(C) Optical mixture.

(D) Stereoisomer mixture.

(17) Which of the following compounds is most likely to display optical activity?

(A) A molecule with a plane of symmetry.

(B) A molecule containing a chiral center and no internal mirror plane.

(C) A meso compound with two chiral centers.

(D) An achiral compound with no stereocenters.

(18) Why do enantiomers have drastically different biological effects in biological systems?

(A) Due to their interactions with achiral environments.

(B) Because of their mirror-image relationship.

(C) Due to their identical physical properties.

(D) Because of their chiral nature and interactions with chiral biological systems.

(19) Which of the following statements regarding the distinguishing features of diastereomers is correct?

(A) They have a mirror-image relationship.

(B) They have identical physical and chemical properties.

(C) They are not mirror images of each other.

(D) They have similar interactions with plane-polarized light.

(20) In what kind of molecules do diastereomers mainly arise?

(A) Molecules with a single chiral center.

(B) Molecules with no chiral centers.

(C) Molecules with multiple chiral centers.

(D) Molecules with symmetrical structures.

(21) Why are primary substrates generally unreactive in SN1 reactions?

(A) They form stable carbocations.

(B) They cannot form carbocations.

(C) They form unstable primary carbocations.

(D) They require strong nucleophiles.

(22) What type of nucleophile is usually sufficient for an SN1 reaction?

(A) Strong.

(B) Weak.

(C) Charged.

(D) Sterically hindered.

(23) Which type of solvent is most favorable for an SN1 reaction?

(A) Polar protic solvents.

(B) Nonpolar solvents.

(C) Polar aprotic solvents.

(D) Any solvent will suffice.

(24) Which of the following is the correct representation of the rate law for an E2 reaction?

(A) Rate = k[substrate].

(B) Rate = k[base].

(C) Rate = k[substrate][base].

(D) Rate = k[substrate]2.

(25) What arrangement of atoms between the hydrogen being removed and the leaving group is ideal for the E2 reaction to occur?

(A) Syn-periplanar.

(B) Anti-periplanar.

(C) Gauche.

(D) Coplanar.

(26) Which reagent is commonly used as an electrophile in the hydroboration step of the hydroboration-oxidation of alkenes?

(A) H_2O.

(B) H_2O_2.

(C) NaOH.

(D) BH_3 or B_2H_6.

(27) Which carbon does the boron atom add to during the hydroboration step of the hydroboration-oxidation of alkenes?

(A) The less substituted carbon.

(B) The more substituted carbon.

(C) Both carbons equally.

(D) It will attach to a hydrogen.

(28) What type of addition occurs during the hydroboration step of the hydroboration-oxidation of alkenes?

(A) Syn addition.

(B) Anti addition.

(C) Elimination.

(D) Substitution.

(29) What happens to the boron atom during the oxidation step of the hydroboration-oxidation of alkenes?

(A) It is eliminated.

(B) It is replaced by a hydroxyl group.

(C) It remains unchanged.

(D) It forms a double bond.

(30) What is the final product after hydroboration-oxidation of an alkene?

(A) Alkane.

(B) Ether.

(C) Alcohol.

(D) Alkyl halide.

(31) How do resonance effects influence vibrational frequencies?

(A) They always increase frequencies.

(B) They can either raise or lower frequencies depending on the bond order.

(C) They have no effect on vibrational frequencies.

(D) They only affect C–H stretching.

(32) Which of the following statements regarding the effect of electron-withdrawing groups on bond strength is correct?

(A) They decrease bond strength.

(B) They have no effect on bond strength.

(C) They can increase bond strength and vibrational frequency.

(D) They only affect resonance structures.

(33) Which of the following bonds would vibrate at the highest frequency?

(A) C–C.

(B) C=C.

(C) C≡C.

(D) All will vibrate at the same frequency.

(34) What does Nuclear Magnetic Resonance (NMR) spectroscopy utilize to provide information about atomic nuclei?

(A) Thermal properties.

(B) Magnetic properties.

(C) Electrical properties.

(D) Optical properties.

(35) Which of the following nuclei are commonly utilized in NMR spectroscopy?

(A) ^{7}N and ^{15}N.

(B) ^{1}H and ^{13}C.

(C) ^{18}O and ^{2}H.

(D) ^{2}H and ^{7}Li.

(36) What effect does temperature have on regioselectivity in the halogenation of alkanes?

(A) Higher temperatures enhance selectivity.

(B) Higher temperatures reduce selectivity.

(C) Temperature has no effect.

(D) Lower temperatures reduce reactivity.

(37) Which of the following is the most common occurrence during the halogenation of an alkane at a stereocenter?

(A) Only one enantiomer is produced.

(B) No reaction takes place.

(C) Racemization occurs.

(D) The product is always optically active.

(38) What complicates the purification of products formed by radical halogenation of alkanes?

(A) High regioselectivity.

(B) Formation of a single product.

(C) Polysubstitution.

(D) Instantaneous reactions.

(39) What is generated during the initiation step of the radical additions to alkenes?

(A) A stable compound.

(B) A radical initiator (X^{-}).

(C) An alkene.

(D) A diene.

(40) What is formed during the initial stage of the propagation step in radical addition to alkenes?

(A) A new carbon-centered radical.

(B) A radical initiator.

(C) Two stable products.

(D) A radical that initiates termination.

(41) What is the active electrophile during the sulfonation process of an aromatic ring?

(A) SO_3.

(B) H_2SO_4.

(C) H_2O.

(D) HSO_4^-.

(42) What intermediate is formed when sulfur trioxide attacks the aromatic ring during the sulfonation process of an aromatic ring?

(A) Arenium ion.

(B) Carbocation.

(C) Sulfonic acid.

(D) Protonated aromatic compound.

(43) Which acid catalyst is most commonly used in Friedel-Crafts alkylation?

(A) Sodium chloride.

(B) Sulfuric acid.

(C) Phosphoric acid.

(D) Aluminum chloride.

(44) What happens when the Lewis acid interacts with the alkyl halide in Friedel-Crafts alkylation?

(A) It decomposes the alkyl halide.

(B) It forms a stable product.

(C) It generates a carbocation or polarized complex.

(D) It reduces the alkyl halide.

(45) What type of shifts can form undesired isomers during Friedel-Crafts alkylation?

(A) 1,2-hydride or alkyl shifts.

(B) 1,2-methyl shifts.

(C) 1,2-ethyl shifts.

(D) 1,2-phenyl shifts.

(46) Which of the following is a characteristic feature of gem-diols?

(A) They are stable in all conditions.

(B) They have both hydroxyl groups on different carbons.

(C) They are unstable and can revert to carbonyl compounds.

(D) They can only be formed from ketones.

(47) What happens to the hydroxyl group of the hemiacetal during acetal formation?

(A) It is eliminated as water.

(B) It remains unchanged.

(C) It is converted into a carbonyl group.

(D) It is oxidized.

(48) Under which of the following conditions are acetals and ketals stable?

(A) Acidic conditions.

(B) Neutral or basic conditions.

(C) High temperatures.

(D) Reducing conditions.

(49) Which of the following products is formed when ketones react with alcohols in the presence of an acid catalyst?

(A) Ketals.

(B) Acetals.

(C) Gem-diols.

(D) Aldehydes.

(50) What type of amine reacts with aldehydes and ketones to form imines?

(A) Primary amines.

(B) Secondary amines.

(C) Tertiary amines.

(D) Quaternary amines.

(51) What type of base is commonly used to form kinetic enolates?

(A) Sodium hydroxide (NaOH).

(B) Sodium ethoxide (NaOEt).

(C) Lithium diisopropylamide (LDA).

(D) Potassium carbonate (K_2CO_3).

(52) Which enolate is favored when immediate reactivity is desired?

(A) Thermodynamic enolate.

(B) Kinetic enolate.

(C) Both enolates are equally favored.

(D) Neither enolate is preferred.

(53) What kind of conditions cause the formation of thermodynamic enolates?

(A) Rapid, irreversible deprotonation.

(B) Use of a strong, bulky base.

(C) Low temperatures with a weak base.

(D) A weak base at higher temperatures.

(54) Which base is usually used to form thermodynamic enolates?

(A) Lithium diisopropylamide (LDA).

(B) Sodium ethoxide (NaOEt).

(C) Potassium hydroxide (KOH).

(D) Sodium bicarbonate ($NaHCO_3$).

(55) What is the main goal in the formation of a thermodynamic enolate?

(A) Achieve a rapid reaction.

(B) Produce a less stable product.

(C) Produce a more stable and substituted enolate.

(D) Minimize reaction time.

(56) What forms as a result of the $KMnO_4$ oxidation of alkenes under basic conditions?

(A) Ketones.

(B) Carboxylic acids.

(C) Syn-diols.

(D) Aldehydes.

(57) What is the first step in the mechanism of oxidation with $KMnO_4$?

(A) Formation of a cyclic permanganate ester intermediate.

(B) Decomposition to manganese dioxide.

(C) Protonation of the alcohol.

(D) Elimination of water.

(58) What is the significant limitation of using $KMnO_4$ as an oxidizing agent?

(A) It is ineffective for most organic compounds.

(B) It can cause over-oxidation and unwanted side reactions.

(C) It only works in non-aqueous solutions.

(D) It is non-toxic and safe to handle.

(59) What is the main function of reducing agents in organic chemistry?

(A) Donate electrons.

(B) Donate protons.

(C) Accept electrons.

(D) Form covalent bonds.

(60) What is the limitation of using $LiAlH_4$ as a reducing agent?

(A) It is ineffective in reducing carbonyl compounds.

(B) It is a strong reducing agent that may cause over-reduction.

(C) It is non-reactive with protic solvents.

(D) It is inexpensive and easy to handle.

(61) What is the benefit of identifying key reactions in retrosynthetic analysis?

(A) They can generate multiple bonds or stereocenters in a single step.

(B) They always produce a single product.

(C) They require more reagents.

(D) They complicate the synthesis process.

(62) What is a guiding principle regarding simplicity in retrosynthetic analysis?

(A) Aim for the most complex synthesis possible.

(B) Avoid the use of any reagents.

(C) Focus on the shortest and most efficient synthesis possible.

(D) Incorporate as many steps as necessary.

(63) Which of the following is included in a realistic approach in retrosynthetic analysis?

(A) Ignore the cost of reagents.

(B) Consider the practicality of each step.

(C) Focus only on theoretical aspects.

(D) Attempt the most challenging reactions available.

(64) Which of the following is the correct outcome of the Diels-Alder Reaction?

(A) Formation of an alcohol.

(B) Formation of an alkene.

(C) Cycloaddition forming a cyclohexene.

(D) Reduction of a carbonyl compound.

(65) Which transformation occurs when an ester undergoes hydrolysis?

(A) Ester to alcohol.

(B) Ester to carboxylic acid.

(C) Ester to aldehyde.

(D) Ester to alkane.

(66) Which of the following statements regarding the effect of crystallinity on polymers is correct?

(A) Crystalline polymers are weaker and more flexible.

(B) Amorphous polymers are stronger and more rigid.

(C) Crystalline polymers tend to be stronger and more rigid.

(D) Crystallinity has no effect on polymer properties.

(67) Which of the following statements regarding thermoplastic polymers is correct?

(A) They undergo irreversible changes upon heating.

(B) They can be repeatedly softened by heating and solidified by cooling.

(C) They exhibit rubber-like elasticity.

(D) They are only found in natural forms.

(68) Which of the following is an example of a thermoset polymer?

(A) Epoxy resin.

(B) Polyethylene.

(C) Natural rubber.

(D) Polyvinyl chloride (PVC).

(69) Which of the following is a defining feature of elastomers?

(A) They form a rigid, cross-linked network.

(B) They can be melted and reshaped multiple times.

(C) They exhibit rubber-like elasticity.

(D) They are only used in industrial applications.

(70) Which polymer is commonly used for plastic bags, films, and bottles?

(A) Polypropylene.

(B) Polystyrene.

(C) Polyethylene.

(D) PVC.

Test 3: Answers and Explanations

(1) (B) C_nH_{2n}.

Cycloalkanes are characterized by the formula C_nH_{2n} due to their ring structure, which introduces a unit of unsaturation. The other formulas apply to different categories of hydrocarbons.

(2) (B) They are subject to angle strain due to their bond angles.

Cyclopropane and cyclobutane experience decreased stability due to angle strain, as their bond angles deviate from the ideal tetrahedral angle of 109.5°. In contrast, cyclohexane exhibits bond angles that are closer to the ideal, particularly in its chair conformation.

(3) (B) The chair conformation.

The chair conformation of cyclohexane is the most stable due to its ability to minimize both steric and torsional strain. This results in bond angles that are close to the ideal value. Other conformations, such as boat and half-chair, exhibit higher energy and increased strain.

(4) (B) The triple bond in alkynes increases the strength of intermolecular forces.

Alkynes, with triple bonds, usually exhibit stronger intermolecular forces than alkenes, which have double bonds. This is due to the linear geometry of alkynes, which allows for closer packing and stronger London dispersion forces compared to the more bent structure of alkenes. Even though both alkynes and alkenes have lower boiling points than their saturated counterparts (alkanes), alkynes generally have higher boiling points than alkenes of comparable molecular weight because the stronger intermolecular forces in alkynes result in more energy required to overcome these forces during phase changes.

(5) (B) Hex-2-ene.

The proper IUPAC name for an alkene with a double bond between carbons 2 and 3 is hex-2-ene. This designation gives the double bond the lowest possible locant. The other options incorrectly assign the double bond to different positions.

(6) (B) It stabilizes the conjugate base and enhances acidity.

In phenolic compounds, when the hydroxyl group (–OH) donates a proton (H^+), the resonance can stabilize the resulting conjugate base (phenoxide ion). The negative charge on the oxygen can be delocalized across the aromatic ring, which stabilizes the conjugate base. This stabilization makes it easier for the phenolic compound to lose a proton, which enhances its acidity compared to alcohols that do not have such resonance stabilization.

(7) (B) Move sigma (σ) bonded atoms or alter σ bonds.

When drawing resonance structures, the focus is on the movement of π electrons or lone pairs, while σ bonds and the positions of atoms must remain unchanged. It is a mistake to move σ-bonded atoms or alter σ bonds, as these bonds are generally not involved in resonance.

(8) (B) The molecule must possess a planar structure.

For a molecule to be considered aromatic, it must be a cyclic, planar, conjugated system that contains $(4n + 2)$ π electrons. Aromatic compounds are inherently cyclic, and conjugation is a fundamental requirement for aromaticity.

(9) (B) Six.

Furan is a five-membered heterocyclic aromatic compound and contains six π electrons (four from two double bonds, plus two from the lone pair on oxygen). It thereby adheres to Hückel's rule of $(4n + 2)$, where n equals 1.

(10) (B) It lacks full conjugation due to structural constraints.

Cyclopentadiene is indeed a cyclic compound. However, it is not fully conjugated because it contains an sp^3-hybridized carbon that disrupts the complete delocalization of π electrons, which prevents it from being aromatic. It is also cyclic and has only four π electrons.

(11) (A) $pK_a + pK_\beta = pK_h = 14$.

For a conjugate acid-base pair in aqueous solution, there is a useful relationship between their pK_a and pK_β values:

- $pK_a + pK_\beta = pK_h = 14$ (at 25°C)

(12) (C) The conjugate base of a weaker acid is less stable and more willing to accept protons.

A weaker acid does not readily donate protons, meaning its conjugate base is relatively strong, as it has a greater tendency to accept protons. This higher proton affinity makes the conjugate base a stronger base, reflecting the inverse relationship between acid and base strength.

(13) (B) Stronger acid.

A lower pK_a value corresponds to a higher acidity or a stronger acid. This is because pK_a is the negative logarithm of the acid dissociation constant (Ka), which measures the strength of an acid in solution.

(14) (A) Acidity increases as electronegativity increases.

For atoms within the same period of the periodic table, acidity generally increases with higher electronegativity. A more electronegative atom can better stabilize the negative charge of the conjugate base, which enhances the acid strength.

(15) (D) Rotation of plane-polarized light in opposite directions.

Enantiomers rotate plane-polarized light in opposite directions. One enantiomer rotates light to the right (dextrorotatory, denoted as (+)), while the other rotates it to the left (levorotatory, denoted as (–)). This property is why enantiomers are sometimes referred to as optical isomers.

(16) (A) Racemic mixture.

A 50:50 mixture of enantiomers (a racemic mixture) does not rotate plane-polarized light because the rotations cancel out. As a result, there will not be a net rotation.

(17) (B) A molecule containing a chiral center and no internal mirror plane.

Optical activity arises when a molecule is chiral, meaning it lacks a plane of symmetry and cannot be superimposed on its mirror image. Such molecules can rotate plane-polarized light, a defining feature of optical activity.

(18) (D) Because of their chiral nature and interactions with chiral biological systems.

Enantiomers can have drastically different biological effects due to the chiral nature of biological systems. For example, the drug thalidomide has two enantiomers: one is effective for morning sickness, while the other causes severe congenital disabilities. Similarly, the (S)-enantiomer of ibuprofen is more pharmacologically active than the (R)-enantiomer.

(19) (C) They are not mirror images of each other.

Diastereomers are stereoisomers that are not mirror images of each other. They arise in molecules with multiple chiral centers or systems with restricted rotation (e.g., double bonds or cyclic structures). Diastereomers exhibit distinct physical and chemical properties, such as boiling points and solubilities. This makes them important in fields such as pharmaceuticals, where their distinct characteristics can impact drug efficacy and interactions.

(20) (C) Molecules with multiple chiral centers.

Diastereomers often occur in molecules with two or more chiral centers, where the stereoisomers differ in the configuration at one or more (but not all) chiral centers. For a molecule with n chiral centers, the maximum number of stereoisomers is 2^n, which may include both enantiomers and diastereomers.

(21) (C) They form unstable primary carbocations.

SN1 is favored by tertiary and resonance-stabilized secondary alkyl halides that can form stable carbocations. Primary substrates are generally unreactive due to unstable primary carbocations.

(22) (B) Weak.

A weak nucleophile is often sufficient in SN1 reactions since the nucleophile does not participate in the rate-determining step, although increased concentration can enhance the reaction rate.

(23) (A) Polar protic solvents.

Polar protic solvents (e.g., water, alcohols) stabilize the carbocation intermediate through solvation. Additionally, a good leaving group is necessary for carbocation formation.

(24) (C) Rate = k[substrate][base].

The rate of an E2 reaction is second order, dependent on both substrate and base concentrations: Rate = k[substrate][base]

(25) (B) Anti-periplanar.

The E2 reaction is a one-step process in which a strong base removes a proton from a carbon adjacent to the one bearing the leaving group. This simultaneously causes the leaving group to depart and form a π bond (alkene). Effective overlap of orbitals during bond breaking and forming is critical. It is ideally achieved when the removed hydrogen and the leaving group are in an anti-periplanar arrangement (dihedral angle of 180°).

(26) (D) BH_3 or B_2H_6.

Hydroboration-oxidation is a two-step process that results in the anti-Markovnikov addition of water across an alkene and is stereospecific, which involves syn addition. During the hydroboration stage, borane (BH_3), often used as its dimer diborane (B_2H_6) or as a borane-Lewis base complex (e.g., BH_3·THF), acts as an electrophile.

(27) (A) The less substituted carbon.

During the hydroboration step of the hydroboration-oxidation of alkenes, the boron atom adds to the less substituted carbon of the alkene. This is due to steric factors and the partial positive charge on the more substituted carbon during the transition state.

(28) (A) Syn addition.

During the hydroboration step of the hydroboration-oxidation of alkenes, the hydrogen atom from BH_3 adds to the more substituted carbon, which results in syn addition (both boron and hydrogen add to the same face of the alkene). This syn addition forms an organoborane intermediate, which can subsequently undergo oxidation to yield an alcohol.

(29) (B) It is replaced by a hydroxyl group.

During the oxidation stage, the trialkylborane intermediate is treated with hydrogen peroxide (H_2O_2) in a basic solution (e.g., NaOH). The boron atom is replaced by a hydroxyl group (OH), which maintains the stereochemistry established during hydroboration.

(30) (C) Alcohol.

The final product of the hydroboration-oxidation process is an alcohol, specifically an alcohol that is formed through anti-Markovnikov addition. This means that the hydroxyl group ends up on the less substituted carbon of the original alkene.

(31) (B) They can either raise or lower frequencies depending on the bond order.

Electron-withdrawing groups can increase bond strength and vibrational frequency. Resonance effects can either raise or lower frequencies depending on the bond order.

(32) (C) They can increase bond strength and vibrational frequency.

Electron-withdrawing groups can enhance the bond strength of adjacent bonds by stabilizing negative charges or increasing the effective nuclear charge on the atoms involved. This stabilization can result in stronger bonds, which vibrate at higher frequencies in IR spectroscopy.

(33) (C) C≡C.

A triple bond (C≡C) vibrates at a higher frequency than a double bond (C=C) or a single bond (C–C). This is due to the increased bond strength and shorter bond length associated with triple bonds, which require more energy to stretch.

(34) (B) Magnetic properties.

NMR spectroscopy utilizes the magnetic properties of atomic nuclei. Nuclei with an odd number of protons or neutrons exhibit nuclear spin, which generates a magnetic moment. When placed in an external magnetic field, these nuclei can align with or against the field. Absorption of radiofrequency radiation can induce a flip in the spin state, and the frequencies at which this occurs provide insight into the magnetic environment of the nuclei within the molecule.

(35) (B) ^{1}H and ^{13}C.

^{1}H and ^{13}C nuclei are mainly used in NMR spectroscopy due to their magnetic properties and abundance in organic compounds. They provide valuable structural information.

(36) (B) Higher temperatures reduce selectivity.

Several factors beyond radical stability influence regioselectivity. Temperature affects selectivity, as higher temperatures provide energy to overcome transition state differences, which reduces selectivity by making all positions more reactive. In contrast, lower temperatures enhance selectivity, particularly for bromination.

(37) (C) Racemization occurs.

When an alkane substrate possesses chirality, radical halogenation at a stereocenter can result in racemization. This occurs because the intermediate alkyl radical formed during the reaction is planar or undergoes rapid inversion, which enables the halogen atom to attack from either side of the molecule.

(38) (C) Polysubstitution.

In polysubstitution, multiple hydrogen atoms in the alkane can be replaced by halogens. This results in a complex mixture of products, each with varying degrees of substitution. The presence of multiple reactive sites increases the likelihood of forming different halogenated products, which makes it difficult to isolate the desired compound.

(39) (B) A radical initiator (X^-).

A radical initiator, which starts the radical chain process, is generated during the initiation step. This initiator can be formed from various sources, such as heat or light, and breaks down into reactive radicals.

(40) (A) A new carbon-centered radical.

In the initial stage of propagation, the radical initiator adds to the π bond of the alkene, which forms a new carbon-centered radical.

(41) (A) SO_3.

Sulfur trioxide (SO_3) is the active electrophile. It can be formed from the equilibrium of sulfuric acid:

$$2H_2SO_4 \rightleftharpoons SO_3 + H_3O^+ + HSO_4^-$$

(42) (A) Arenium ion.

During the electrophilic attack stage, sulfur trioxide attacks the aromatic ring, which forms the arenium ion intermediate. Next, a proton transfer occurs, which yields benzenesulfonic acid during the proton transfer stage.

(43) (D) Aluminum chloride.

Friedel-Crafts alkylation introduces an alkyl group onto the aromatic ring using an alkyl halide (R–X) in the presence of a Lewis acid catalyst, most commonly aluminum chloride ($AlCl_3$). The Lewis acid helps generate a carbocation or polarizes the alkyl halide, which makes the carbon atom electrophilic.

(44) (C) It generates a carbocation or polarized complex.

When the Lewis acid catalyst (e.g., $AlCl_3$) interacts with the alkyl halide (R–X), it forms a carbocation or a polarized complex:

$R{-}X + AlCl_3 \rightleftharpoons R^+ [AlCl_4]^-$ (or $R\delta^+$---$X\delta^-$---$AlCl_3$)

(45) (A) 1,2-hydride or alkyl shifts.

If the alkyl halide can form a primary or secondary carbocation, it may rearrange to a more stable carbocation (secondary or tertiary) through 1,2-hydride or alkyl shifts, which results in a mixture of alkylated products. This is a significant limitation, as it can cause the formation of undesired isomers.

(46) (C) They are unstable and can revert to carbonyl compounds.

Gem-diols are generally unstable and readily revert to the carbonyl compound by losing water. However, in some cases, such as formaldehyde, the gem-diol is the predominant form in aqueous solution.

(47) (A) It is eliminated as water.

The hydroxyl group of the hemiacetal is protonated, which converts it into a good leaving group (water). Water is eliminated, which generates an oxonium ion. Then, another alcohol molecule attacks the oxonium ion, which forms the acetal (or ketal).

(48) (B) Neutral or basic conditions.

Acetals and ketals are stable under neutral or basic conditions but are readily hydrolyzed back to the carbonyl compound and alcohol under acidic conditions. Acetals and ketals are often used as protecting groups for aldehydes and ketones, as they can be easily introduced and removed under mild conditions.

(49) (A) Ketals.

Aldehydes and ketones react with alcohols in the presence of an acid catalyst to form acetals (from aldehydes) or ketals (from ketones). This reaction involves two successive nucleophilic additions of alcohol molecules.

(50) (A) Primary amines.

Aldehydes and ketones react with primary amines (RNH_2) to form imines (i.e., Schiff bases). This reaction involves the nucleophilic addition of the amine to the carbonyl group, followed by elimination of water.

(51) (C) Lithium diisopropylamide (LDA).

The kinetic enolate is formed under conditions that favor rapid, irreversible deprotonation. Usually, this involves the use of a strong, bulky base, such as LDA, at low temperatures.

(52) (B) Kinetic enolate.

The kinetic enolate is often the preferred product when immediate reactivity is desired, such as in specific synthetic routes. This is because the pathway to form the kinetic enolate is less hindered by steric factors, which allows for a faster reaction rate.

(53) (D) A weak base at higher temperatures.

In contrast, the thermodynamic enolate is formed under conditions that allow for equilibrium to be established. Under these conditions, the reaction can proceed to completion. This allows for the more stable, substituted enolate to form.

(54) (B) Sodium ethoxide (NaOEt).

The process usually employs a weaker base, such as sodium ethoxide (NaOEt), at higher temperatures. Thermodynamic enolates are formed under conditions that allow the reaction to reach equilibrium.

(55) (C) Produce a more stable and substituted enolate.

The thermodynamic enolate is usually favored when the goal is to produce a more stable product, as the stability of the enolate is enhanced by greater substitution at the α-carbon.

(56) (C) Syn-diols.

Potassium permanganate ($KMnO_4$) is a strong oxidizing agent that can oxidize a wide variety of organic compounds. It is usually used in aqueous solution, either under acidic or basic conditions.

- Acidic conditions: $KMnO_4$ oxidizes primary alcohols to carboxylic acids and secondary alcohols to ketones.
- Basic conditions: $KMnO_4$ can oxidize alkenes to syn-diols (dihydroxylation).

(57) (A) Formation of a cyclic permanganate ester intermediate.

The mechanism of oxidation with $KMnO_4$ is complex and depends on the reaction conditions. In general, it involves the formation of a cyclic permanganate ester intermediate, followed by decomposition to the oxidized product and manganese dioxide (MnO_2).

(58) (B) It can cause over-oxidation and unwanted side reactions.

The limitations of $KMnO_4$ include the following.

- Strong oxidizing agent: It can cause over-oxidation and unwanted side reactions.
- Harsh conditions: This issue can limit its use with sensitive compounds.
- Formation of MnO_2: This byproduct can make product isolation difficult.
- Not suitable for acid-sensitive compounds: Acidic conditions are often employed, which can be problematic for these compounds.

(59) (A) Donate electrons.

Reducing agents are substances that donate electrons, which cause another species to be reduced. In organic chemistry, reduction usually increases the number of bonds to hydrogen or decreases the number of bonds to oxygen. Several reducing agents are commonly used to reduce carbonyl compounds, carboxylic acids, and other functional groups.

(60) (B) It is a strong reducing agent that may result in over-reduction.

The following are the limitations of $LiAlH_4$.

- Strong reducing agent: It can reduce a wide variety of functional groups, which can impact selectivity.
- Highly reactive: It must be handled with care.
- Reacts with protic solvents: It reacts violently with water and other protic solvents.
- Expensive: It is a relatively costly reagent.

(61) (A) They can generate multiple bonds or stereocenters in a single step.

Identifying key reactions that can generate multiple bonds or stereocenters in a single step is an important aspect of retrosynthetic analysis. These reactions enable the rapid construction of complex molecular architectures, which can significantly streamline the synthesis process. By leveraging powerful reactions, such as the Robinson annulation or Michael addition, chemists can efficiently build the desired structure while minimizing the number of synthetic steps required.

(62) (C) Focus on the shortest and most efficient synthesis possible.

Simplicity is a guiding principle in retrosynthetic analysis. Chemists should strive for the shortest and most efficient synthesis possible, as this avoids unnecessary steps and reagents. A straightforward synthetic route not only reduces the overall time and effort required for synthesis but also minimizes the potential for errors and side reactions. By prioritizing simplicity, chemists can enhance the feasibility of their synthetic plans and improve the likelihood of successful outcomes.

(63) (B) Consider the practicality of each step.

It is essential to approach retrosynthetic analysis with a realistic mindset. This involves considering the practicality of each step in the proposed synthesis, which includes the cost of reagents, the availability of starting materials, and the potential for side reactions. By evaluating these factors, chemists can develop a more accurate and achievable synthetic plan. A realistic approach ensures that the proposed synthesis is not only theoretically sound but also practically feasible, which ultimately yields successful experimental outcomes.

(64) (C) Cycloaddition forming a cyclohexene.

The Diels-Alder reaction is a [4+2] cycloaddition reaction that involves a diene and a dienophile that forms a six-membered ring. This reaction specifically yields a cyclohexene structure. It is a key reaction in organic synthesis for constructing cyclic compounds, characterized by the formation of new carbon-carbon bonds.

(65) (B) Ester to carboxylic acid.

When an ester undergoes hydrolysis, it reacts with water (either under acidic or basic conditions) to yield a carboxylic acid and an alcohol. This reaction essentially breaks the ester bond and adds a water molecule, which results in the formation of the carboxylic acid and the corresponding alcohol.

(66) (C) Crystalline polymers tend to be stronger and more rigid.

The properties of a polymer depend on its chemical structure, molecular weight, and morphology (the arrangement of the polymer chains). Some important properties of polymers include:

- Tensile strength: The ability of a polymer to resist stretching or breaking.
- Elasticity: The ability of a polymer to return to its original shape after being deformed.

- Glass transition temperature (Tg): The temperature at which a polymer transitions from a rigid, glassy state to a rubbery state.
- Crystallinity: The degree to which the polymer chains are ordered in a crystalline structure. Crystalline polymers tend to be stronger and more rigid than amorphous polymers.

(67) (B) They can be repeatedly softened by heating and solidified by cooling.

Thermoplastics are a type of polymer that can be repeatedly softened by heating and solidified by cooling. Examples include polyethylene, polypropylene, polystyrene, and PVC.

(68) (A) Epoxy resin.

Thermosets are a type of polymer that undergo irreversible chemical changes upon heating, which creates a rigid, cross-linked network. Examples include epoxy resins, phenolic resins, and vulcanized rubber.

(69) (C) They exhibit rubber-like elasticity.

Elastomers are a type of polymer that exhibits rubber-like elasticity. Examples include natural rubber, synthetic rubber (e.g., styrene-butadiene rubber, SBR), and silicone rubber.

(70) (C) Polyethylene.

In the packaging industry, polymers such as polyethylene (PE), polypropylene (PP), and polystyrene (PS) are dominant due to their low cost and adaptability. PE is available in low-density (LDPE) and high-density (HDPE) forms and used for plastic bags, films, and bottles. It provides barrier properties against moisture, gases, and UV light, which extends the shelf life of food, beverages, and other products. Its ability to be molded into various shapes and compatibility with printing for branding further enhance its utility.

Test 4: Questions

(1) What is the reason that alkenes exhibit greater reactivity compared to alkanes?

(A) They contain a triple bond that enhances reactivity.

(B) They are considered saturated compounds.

(C) They possess a double bond that is more reactive.

(D) They have stronger intermolecular forces than alkanes.

(2) What is the correct IUPAC name for a compound with a five-carbon chain containing both a double bond between carbons 1 and 2 and a triple bond between carbons 4 and 5?

(A) Pent-1-en-4-yne.

(B) Pent-4-en-1-yne.

(C) Pent-2-en-4-yne.

(D) Pent-1-yne-4-ene.

(3) In naming a branched alkane like 3-methylhexane, how is the parent chain numbered?

(A) Give the highest number to the methyl group.

(B) Give the lowest number to the methyl group.

(C) Maximize the number of substituents.

(D) Include the functional group.

(4) For 2,3-dimethylpent-2-ene, which stereochemical descriptor is necessary due to potential geometric isomerism?

(A) R/S descriptors are needed.

(B) E/Z descriptors are needed.

(C) No descriptors are needed.

(D) Cis/trans descriptors are optional.

(5) What prefix is utilized for an ether functional group when it does not serve as the principal functional group in a compound?

(A) Hydroxy-.

(B) Alkoxy-.

(C) Oxo-.

(D) Amino-.

(6) What characterizes an anti-aromatic compound?

(A) $4n + 2$ π electrons.

(B) Non-planar geometry.

(C) $4n$ π electrons.

(D) Lack of conjugation.

(7) Which compound is anti-aromatic?

(A) Benzene.

(B) Cyclopentadienyl anion.

(C) Cyclobutadiene.

(D) Pyrrole.

(8) Why are triple bonds stronger than double bonds?

(A) They contain more sigma (σ) bonds than double bonds do.

(B) They consist of an increased number of π bonds compared to double bonds.

(C) They exhibit longer bond lengths than double bonds.

(D) They possess less s-character in their hybrid orbitals.

(9) How does sp^2 hybridization affect molecular geometry?

(A) Creates a tetrahedral shape.

(B) Forms a linear structure.

(C) Produces a trigonal planar shape.

(D) Results in a bent geometry.

(10) What determines a molecule's polarity?

(A) Hybridization and bond polarity.

(B) Bond strength alone.

(C) Number of σ bonds.

(D) Resonance structures only.

(11) Why does basicity usually decrease with increasing atomic size in the same group of the periodic table?

(A) Larger atoms have stronger electron-donating groups.

(B) Larger atoms have more electron-withdrawing groups.

(C) Larger atoms have more diffuse lone pairs, which makes them less available to accept a proton.

(D) Larger atoms have less polarizability.

(12) How does resonance stabilization impact the acidity of a compound?

(A) Increases acidity.

(B) Decreases acidity.

(C) Does not affect acidity.

(D) Increases basicity.

(13) How do electron-withdrawing groups affect the acidity of a compound?

(A) Increases the acidity by pulling electron density away from the negative charge.

(B) Decreases the acidity by donating electrons to the negative charge.

(C) Has no effect on acidity.

(D) Increases the basicity by stabilizing the conjugate base.

(14) What effect do electron-donating groups near a basic site have on basicity?

(A) Decreases the basicity by enhancing the ability to accept a proton.

(B) Increases the basicity by increasing electron density on the basic atom and enhancing its ability to accept a proton.

(C) Has no effect on basicity.

(D) Increases the acidity by pulling electron density toward the negative charge.

(15) What term is used to describe a chiral compound that rotates plane-polarized light clockwise?

(A) Dextrorotatory.

(B) Levorotatory.

(C) Racemic.

(D) Isomeric.

(16) All of the following factors influence the extent of rotation of plane-polarized light by a chiral compound except:

(A) Concentration.

(B) Temperature.

(C) Wavelength of light.

(D) Acidity of the compound.

(17) What would be the net optical rotation of a racemic mixture?

(A) 0.

(B) +20.

(C) −20.

(D) It varies depending on the compound.

(18) What is the purpose of the Cahn-Ingold-Prelog priority rules in stereochemistry?

(A) Determine the molecular weight of a compound.

(B) Assign configurations (R or S) to stereocenters.

(C) Analyze the chirality of an achiral compound.

(D) Identify isotopes in a molecule.

(19) How are the substituents attached to a chiral center ranked in the Cahn-Ingold-Prelog priority rules?

(A) Based on their size.

(B) Based on their position in the molecule.

(C) Based on the atomic number of the atoms directly bonded to the chiral center.

(D) Based on their color.

(20) In the R/S nomenclature, what configuration is assigned if the sequence from highest to lowest priority is counterclockwise?

(A) (R).

(B) (S).

(C) (M).

(D) (A).

(21) What type of solvent is most favorable for the E2 (bimolecular elimination) mechanism?

(A) Polar protic solvents.

(B) Nonpolar solvents.

(C) Polar aprotic solvents.

(D) Any solvent will suffice.

(22) What is the rate law for an E1 reaction?

(A) Rate = k[substrate].

(B) Rate = k[base].

(C) Rate = k[substrate][base].

(D) Rate = k[substrate]^2.

(23) Which type of substrates are most favorable for E1 reactions?

(A) Primary alkyl halides.

(B) Secondary alkyl halides.

(C) Tertiary and resonance-stabilized secondary alkyl halides.

(D) All substrates are equally favorable for E1 reactions.

(24) Which type of carbocation is generally unstable and not involved in SN1 or E1 reactions?

(A) Tertiary carbocations.

(B) Secondary carbocations.

(C) Primary carbocations.

(D) Resonance-stabilized carbocations.

(25) What is the reactivity order for SN2 reactions based on substrate structure?

(A) Tertiary > secondary > primary > methyl.

(B) Methyl > primary > secondary > tertiary.

(C) Secondary > primary > methyl > tertiary.

(D) Primary > methyl > secondary > tertiary.

(26) What does regioselectivity refer to in a chemical reaction?

(A) Preference for forming stereoisomers.

(B) Preference for bond formation at a specific atom.

(C) Rate of reaction.

(D) Temperature dependence.

(27) What is the result of a reaction that exhibits anti-Markovnikov regioselectivity?

(A) The major product has the electrophile on the less substituted carbon.

(B) The major product has the electrophile on the more substituted carbon.

(C) No products are formed.

(D) The reaction rate is slower.

(28) Which factor most directly influences regioselectivity in a reaction that involves alkenes?

(A) Temperature.

(B) Electrophile strength.

(C) Carbocation stability.

(D) Concentration of reactants.

(29) What is the significance of carbocation stability in determining regioselectivity?

(A) More stable carbocations result in less favorable products.

(B) More stable carbocations result in more favorable products.

(C) Carbocation stability has no effect.

(D) All carbocations result in less favorable products.

(30) Which of the following reactions exemplifies anti-Markovnikov regioselectivity?

(A) Acid-catalyzed hydration.

(B) Electrophilic addition of HX.

(C) Dehydrohalogenation.

(D) Hydroboration-oxidation.

(31) Which of the following statements regarding the concept of chemical shift in ^1H NMR is correct?

(A) No energy is required to induce a spin flip.

(B) The resonance frequency is affected by electron density.

(C) The number of hydrogen atoms in a molecule affects the frequency.

(D) The distance between peaks in a multiplet affects the shift.

(32) Which of the following is the correct unit to report chemical shifts in NMR spectroscopy?

(A) Parts per million (ppm).

(B) Decibels (dB).

(C) Millimeters (mm).

(D) Hertz (Hz).

(33) What does the area under each peak in a ^1H NMR spectrum indicate?

(A) The number of equivalent protons that contribute to that signal.

(B) The chemical formula of the molecule.

(C) The temperature of the sample.

(D) The energy of the nucleus.

(34) Which of the following statements regarding the significance of the n + 1 rule in ^1H NMR is correct?

(A) It determines chemical shifts.

(B) It predicts the number of peaks in a multiplet.

(C) It calculates integration values.

(D) It measures coupling constants.

(35) Which of the following is the reference compound used to define chemical shifts in ^1H NMR?

(A) Acetic acid.

(B) Water.

(C) Tetramethylsilane (TMS).

(D) Benzene.

(36) What is generated during the initiation step of anti-Markovnikov HBr addition to alkenes?

(A) Stable alkene.

(B) Alkoxy radicals.

(C) A diene.

(D) Hydrogen gas.

(37) What happens initially during the propagation stage of the anti-Markovnikov HBr addition to alkenes?

(A) A stable product is formed.

(B) An alcohol and a bromine radical are formed.

(C) HBr is completely consumed.

(D) No further reactions occur.

(38) What is the outcome of radical initiator addition to a monomer during polymerization?

(A) Generation of new radicals that continue to react until termination.

(B) Formation of small molecules.

(C) Instant termination of the reaction.

(D) Production of a single stable compound.

(39) What challenge is associated with radical polymerization?

(A) Management of heat generation during exothermic reactions.

(B) Lack of product variety.

(C) Excessive use of solvents.

(D) Inability to produce rigid materials.

(40) What is one goal of green chemistry in the context of radical polymerization?

(A) Increase the use of toxic solvents.

(B) Enhance reaction speeds.

(C) Develop water-based or solvent-free processes.

(D) Reduce the variety of polymers produced.

(41) What effect do strong electron-withdrawing groups have on the reactivity of an aromatic compound in Friedel-Crafts reactions?

(A) They enhance reactivity.

(B) They have no effect.

(C) They hinder electrophilic substitution.

(D) They promote multiple substitutions.

(42) Which of the following statements correctly describes Friedel-Crafts acylation?

(A) The introduction of an acyl group onto an aromatic ring.

(B) The introduction of a nitro group onto an aromatic ring.

(C) The introduction of a nitro group onto alkenes.

(D) The introduction of an acyl group onto alkenes.

(43) What is the active electrophile in Friedel-Crafts acylation?

(A) Acyl chloride.

(B) Acylium ion (RCO^+).

(C) Benzene.

(D) Lewis acid.

(44) Why is polyacylation less likely in Friedel-Crafts acylation?

(A) The acylium ion is unstable.

(B) The acyl group is a strong activating group.

(C) The acyl group is a deactivating group.

(D) The reaction does not proceed at all.

(45) What is the major advantage of Friedel-Crafts acylation over Friedel-Crafts alkylation?

(A) It produces more products.

(B) It is faster.

(C) It requires less energy.

(D) It does not result in carbocation rearrangements.

(46) What type of amine reacts with aldehydes and ketones to form enamines?

(A) Secondary amines.

(B) Tertiary amines.

(C) Primary amines.

(D) Quaternary amines.

(47) What is the tetrahedral intermediate formed during imine formation?

(A) Carboxylic acid.

(B) Hemiacetal.

(C) Aldol.

(D) Carbinolamine.

(48) Which of the following functional groups contains a hydroxyl group as its defining feature?

(A) Aldehyde.

(B) Ketone.

(C) Alcohol.

(D) Ester.

(49) What intermediate is formed by the elimination of water in enamine formation?

(A) Carbinolamine.

(B) Iminium ion.

(C) Hemiacetal.

(D) Gem-diol.

(50) Why are enamines considered useful synthetic intermediates?

(A) They are stable and unreactive.

(B) They are easily eliminated.

(C) They do not participate in further reactions.

(D) They can act as nucleophiles in various reactions.

(51) What does it mean for enolates to be ambident nucleophiles?

(A) They can react with multiple electrophiles at once.

(B) They can only react at the α-carbon.

(C) They do not participate in nucleophilic reactions.

(D) They can react at two different sites.

(52) What is the primary product of C-alkylation of an enolate?

(A) An enol ether.

(B) A carbonyl compound.

(C) A carboxylic acid.

(D) A new carbon-carbon bond.

(53) What type of compounds are formed in an aldol reaction?

(A) β-hydroxyaldehydes or β-hydroxyketones.

(B) α,β-unsaturated aldehydes or ketones.

(C) Dicarbonyl compounds.

(D) Esters.

(54) In a crossed aldol reaction, how can the formation of multiple products be minimized?

(A) Use two carbonyl compounds in equal proportions.

(B) Use excess base.

(C) Use high temperature.

(D) Use only one carbonyl compound.

(55) Which products does the Claisen condensation mainly involve?

(A) Two aldehydes.

(B) Two esters.

(C) A ketone and an alcohol.

(D) A carbonyl and a carboxylic acid.

(56) What happens during the nucleophilic attack in the reduction mechanism with $LiAlH_4$?

(A) The alkoxide intermediate is formed.

(B) Water is produced.

(C) A carboxylic acid is formed.

(D) The aluminum atom is reduced.

(57) Which functional groups are commonly reduced by $NaBH_4$?

(A) Carboxylic acids and amides.

(B) Aldehydes and ketones.

(C) Esters and nitriles.

(D) Esters and amides.

(58) Which of the following statements regarding the advantage of $NaBH_4$ is correct?

(A) It is less expensive than $LiAlH_4$.

(B) It can reduce carboxylic acids effectively.

(C) It reacts violently with water.

(D) It requires no solvent.

(59) What happens after the reduction is complete when using $NaBH_4$?

(A) The mixture is heated.

(B) The mixture is filtered.

(C) No further steps are needed.

(D) The reaction mixture is treated with dilute acid to protonate the alkoxide.

(60) All of the following are common catalysts used in catalytic hydrogenation except:

(A) Palladium (Pd).

(B) Platinum (Pt).

(C) Nickel (Ni).

(D) Sodium (Na).

(61) Protecting groups must be easy to introduce in order to:

(A) Complicate the synthesis process.

(B) Ensure it can be attached without extensive procedures.

(C) Increase the cost of the synthesis.

(D) Make the functional group more reactive.

(62) What does it mean for a protecting group to be stable under reaction conditions?

(A) It reacts with all reagents.

(B) It remains intact during the synthesis.

(C) It is removed easily.

(D) It enhances the reactivity of the molecule.

(63) What is the benefit of a protecting group that is easily removable?

(A) Allows for permanent modifications.

(B) Complicates the synthesis process.

(C) Increases the number of steps in the synthesis.

(D) Avoids affecting other functional groups during deprotection.

(64) What role do functional group interconversions (FGI) play in introducing stereocenters?

(A) Introduce new stereocenters in a molecule.

(B) Always prevent stereochemistry.

(C) Have no effect on stereochemistry.

(D) Only affect the molecular weight.

(65) What is the purpose of using silyl ethers as protecting groups for alcohols?

(A) Enhance reactivity.

(B) Increase boiling point.

(C) Form esters.

(D) Prevent unwanted reactions of the alcohol.

(66) What type of packaging incorporates polymer-based sensors to monitor food freshness?

(A) Traditional packaging.

(B) Biodegradable packaging.

(C) Smart packaging.

(D) Rigid packaging.

(67) Which polymer is a key material in construction for pipes and window frames?

(A) Polyurethane (PU).

(B) Polylactic acid (PLA).

(C) PVC.

(D) Acrylonitrile butadiene styrene (ABS).

(68) Which of the following best describes the structure of polystyrene?

(A) A copolymer of styrene and ethylene.

(B) A polymer consisting of repeated benzene rings connected by oxygen atoms.

(C) A chain of repeated styrene units with pendant phenyl groups.

(D) A cyclic polymer formed from benzaldehyde.

(69) Which of the following are common uses of fiber-reinforced polymers (FRPs)?

(A) Food packaging.

(B) Insulation.

(C) Disposable items.

(D) Bridges and structural components.

(70) Which polymer is commonly used to form rigid containers like yogurt cups?

(A) Polyethylene.

(B) Polyurethane.

(C) Polypropylene.

(D) PVC.

Test 4: Answers and Explanations

(1) (C) They possess a double bond that is more reactive.

High reactivity is a key characteristic of alkenes. They contain a carbon-carbon double bond (C=C), which is a region of high electron density. This makes them more reactive than alkanes, which only contain single bonds (C–C). The double bond in alkenes is susceptible to reactions such as addition reactions, where other atoms or groups can add across the double bond.

(2) (A) Pent-1-en-4-yne.

The parent chain is five carbons (pent-), with a double bond at carbon 1 (-ene) and a triple bond at carbon 4 (-yne). The double bond gets the lower number, so the name is pent-1-en-4-yne.

(3) (B) Give the lowest number to the methyl group.

In IUPAC nomenclature for branched alkanes, the parent chain is numbered in such a way as to give the lowest possible numbers to the substituents. This means that the numbering maximizes the number of substituents that receive the lowest numbers.

For example, in 3-methylhexane, the parent chain (hexane) is numbered from the side that gives the methyl group (the substituent) the lowest number, which is three in this case. This approach ensures that the overall structure is clearly defined and the substituents are correctly identified.

(4) (B) E/Z descriptors are needed.

The double bond in 2,3-dimethylpent-2-ene allows geometric isomerism (cis/trans or E/Z) due to restricted rotation. Omitting E/Z descriptors fails to specify which isomer is present and results in ambiguity.

(5) (B) Alkoxy-.

When ethers do not serve as the principal functional group in a compound, they are named using the prefix alkoxy-. This prefix is derived from the alkyl group attached to the oxygen atom. For example, in a compound like ethoxyethane, the prefix ethoxy- indicates the presence of an ethyl group attached to the oxygen, while ethane indicates the principal alkane structure.

(6) (C) $4n$ π electrons.

Anti-aromatic compounds are characterized by having 4n π electrons (where n is an integer). This configuration causes instability and high reactivity. In contrast, aromatic compounds have 4n + 2 π electrons and exhibit enhanced stability due to resonance and planarity. Anti-aromatic compounds do not have this stability and often prefer non-planar conformations to minimize strain.

(7) (C) Cyclobutadiene.

Cyclobutadiene contains four π electrons (4n, where n equals 1), which categorizes it as an anti-aromatic compound. In contrast, benzene, cyclopentadienyl anion, and pyrrole are aromatic due to their respective six π electrons.

(8) (B) They consist of an increased number of π bonds compared to double bonds.

A double bond has one σ bond and one π bond, while a triple bond consists of one σ bond and two π bonds. The additional π bond in a triple bond contributes to greater electron density between the two atoms, which results in a stronger bond compared to a double bond.

(9) (C) Produces a trigonal planar shape.

In sp^2 hybridization, one s orbital mixes with two p orbitals to form three equivalent sp^2-hybrid orbitals. These three sp^2-hybrid orbitals arrange themselves in a trigonal planar geometry, with bond angles of approximately 120 degrees. This arrangement minimizes electron pair repulsion and is characteristic of molecules like ethylene (C_2H_4) and other compounds with sp^2-hybridized carbon atoms.

(10) (A) Hybridization and bond polarity.

The polarity of a molecule is influenced by the following.

- Hybridization: Affects molecular geometry.
- Bond polarity: Arises from differences in electronegativity between bonded atoms.

(11) (C) Larger atoms have more diffuse lone pairs, which makes them less available to accept a proton.

For atoms in the same group of the periodic table, acidity usually increases with larger atomic size. The negative charge of the conjugate base is spread over a greater volume, which causes increased stability. Larger atoms are also more polarizable, which helps stabilize the conjugate base through solvation. For the same group, basicity usually decreases with increasing atomic size. Larger atoms have more diffuse lone pairs, which makes them less available to accept a proton.

(12) (A) Increases acidity.

If the conjugate base can be stabilized by resonance, the acidity of the corresponding acid increases. Delocalization of the negative charge through resonance spreads the charge over multiple atoms, which enhances the stability of the conjugate base and thus increases acid strength. Carboxylic acids are more acidic than alcohols due to resonance stabilization of the carboxylate anion.

(13) (A) Increases the acidity by pulling electron density away from the negative charge.

Electron-withdrawing groups (EWGs) near the acidic proton stabilize the conjugate base by pulling electron density away from the negative charge, which increases acidity. Electronegative atoms (e.g., halogens, oxygen, and nitrogen) and groups with multiple electronegative atoms (e.g., NO_2, CN) serve as EWGs. For example, chloroacetic acid is more acidic than acetic acid due to the electron-withdrawing effect of chlorine.

(14) (B) Increases the basicity by increasing electron density on the basic atom and enhancing its ability to accept a proton.

Electron-donating groups (EDGs) near a basic site increase electron density on the basic atom. This enhances its ability to accept a proton and thus increases basicity. Alkyl groups are weak EDGs. For instance, alkylamines are generally more basic than ammonia. EWGs decrease basicity by pulling electron density away from the basic site.

(15) (A) Dextrorotatory.

Chiral molecules can rotate the plane of plane-polarized light, a property known as optical activity. When plane-polarized light passes through a chiral compound, the plane of polarization is altered. A chiral compound that rotates light clockwise is termed dextrorotatory and is labeled with a (+) or d- prefix. A compound that rotates light counterclockwise is called levorotatory and is designated by a (–) or l- prefix.

(16) (D) Acidity of the compound.

The extent of rotation depends on the compound's concentration, the light's path length through the sample, the wavelength of light, and temperature. The specific rotation [α] quantifies a compound's ability to rotate plane-polarized light: $[\alpha]\lambda T = \alpha / c \cdot l$.

(17) (A) 0.

Enantiomers have equal magnitudes of specific rotation but with opposite signs. A racemic mixture contains equal amounts of both enantiomers. This arrangement results in no net optical rotation, as the effects of the two cancel each other out.

(18) (B) Assign configurations (R or S) to stereocenters.

The Cahn-Ingold-Prelog (CIP) priority rules are employed to assign configurations (R or S) to stereocenters in chiral molecules and provide a clear method to describe the three-dimensional arrangement of substituents.

(19) (C) Based on the atomic number of the atoms directly bonded to the chiral center.

The four substituents attached to the chiral center are ranked based on the atomic number of the atoms directly bonded to it. Higher atomic numbers receive higher priority (1 > 2 > 3 > 4).

(20) (B) (S).

The final stage in CIP priority rules involves examining the order of the remaining three substituents (1, 2, and 3). If the sequence from highest to lowest priority is clockwise, the configuration is (R). If counterclockwise, it is (S).

(21) (C) Polar aprotic solvents.

Polar aprotic solvents favor E2 reactions, as they do not strongly solvate the anionic base. Moreover, a strong base (usually anionic, e.g., OH^-, RO^-, $NaNH_2$) is required, and a good leaving group is essential.

(22) (A) Rate = k[substrate].

The rate law for an E1 reaction is first order and depends only on the substrate concentration: Rate = k[substrate].

(23) (C) Tertiary and resonance-stabilized secondary alkyl halides.

E1 is favored by tertiary and resonance-stabilized secondary alkyl halides capable of forming stable carbocations, similar to SN1. Additionally, E1 reactions favor the formation of the more stable alkene (Zaitsev's rule). This is because the carbocation can lose any adjacent β-proton without requiring a specific coplanar geometry.

(24) (C) Primary carbocations.

Substrates that can form stable carbocations (tertiary or secondary with resonance stabilization) favor SN1 and E1 mechanisms. Primary carbocations are unstable and not usually involved in these reactions.

(25) (B) Methyl > primary > secondary > tertiary.

Branching near the reaction center can hinder backside attack by the nucleophile in SN2 reactions. Usually, the reactivity order for SN2 is methyl > primary > secondary > tertiary. Steric hindrance has less impact on SN1 and E1 reactions, which involve carbocation intermediates.

(26) (B) Preference for bond formation at a specific atom.

Regioselectivity refers to the preference for bond formation at a specific atom within a molecule. In addition, regioselectivity often determines which carbon atom of the multiple bond the electrophile and nucleophile will attach to in reactions involving alkenes and alkynes.

(27) (A) The major product has the electrophile on the less substituted carbon.

Anti-Markovnikov regioselectivity occurs when the electrophile adds to the less substituted carbon atom of an alkene. This is contrary to Markovnikov's rule, which states that the electrophile usually adds to the more substituted carbon.

(28) (C) Carbocation stability.

Carbocation stability helps determine regioselectivity. When a more stable carbocation is generated, it influences where substituents will add to the alkene, which then affects regioselectivity.

(29) (B) More stable carbocations result in more favorable products.

The stability of carbocations is significant because more stable carbocations are lower in energy and are formed more readily during reactions. This stability influences the regioselectivity of the reaction, which results in the production of specific isomers.

(30) (D) Hydroboration-oxidation.

Reactions that proceed with opposite regioselectivity to Markovnikov's rule are termed anti-Markovnikov additions. Hydroboration-oxidation exemplifies this, as hydrogen from BH_3 adds to the more substituted carbon, and the hydroxyl group from oxidation ends up on the less substituted carbon. Radical addition of HBr to alkenes in the presence of peroxides also exhibits anti-Markovnikov regioselectivity.

(31) (B) The resonance frequency is affected by electron density.

Chemical shift refers to how the resonance frequency of protons changes based on their electronic environment. Protons near EWGs resonate at higher frequencies (higher ppm) due to deshielding. Those near EDGs resonate at lower frequencies (lower ppm) due to shielding. This shift provides insight into the molecular structure and functional groups present.

(32) (A) Parts per million (ppm).

The resonance frequency of a proton is affected by the surrounding electron density. EWGs deshield protons and cause them to resonate at higher frequencies (higher ppm values). EDGs shield protons, which results in lower frequencies. Chemical shifts are reported in ppm relative to tetramethylsilane (TMS), defined as 0 ppm.

(33) (A) The number of equivalent protons that contribute to that signal.

The area under each peak in a ^{1}H NMR spectrum correlates with the number of equivalent protons that contribute to that signal. This allows for the determination of the relative amounts of different types of hydrogen atoms.

(34) (B) It predicts the number of peaks in a multiplet.

The magnetic field experienced by a proton can be influenced by the spin of nearby non-equivalent protons, which results in signal splitting into multiple peaks (multiplets). The number of peaks follows the $n + 1$ rule, where n represents the number of equivalent neighboring protons.

(35) (C) Tetramethylsilane (TMS).

TMS is the standard reference compound in NMR spectroscopy because it has a high degree of symmetry, low reactivity, and a single peak in the spectrum. Its chemical shift is defined as 0 ppm, which allows for consistent comparison of shifts in other compounds.

(36) (B) Alkoxy radicals.

During the initiation step, peroxides decompose to form two alkoxy radicals:

$$ROOR \rightarrow 2RO\bullet$$

(37) (B) An alcohol and a bromine radical are formed.

During the initial propagation step, the alkoxy radical reacts with an HBr molecule and produces an alcohol and a bromine radical:

$$RO\bullet + HBr \rightarrow ROH + Br\bullet$$

(38) (A) Generation of new radicals that continue to react until termination.

A radical initiator starts the chain reaction. It adds to a monomer unit and generates a new radical that can then react with additional monomers. It continues until termination occurs.

(39) (A) Management of heat generation during exothermic reactions.

The advantages of radical polymerization include its robustness and compatibility with a variety of monomers. Innovations, such as controlled radical polymerization (e.g., atom transfer radical polymerization), allow for precise control over polymer architecture. This enables the production of advanced materials, such as self-healing coatings. However, it can be difficult to manage heat generation during exothermic reactions and minimize side reactions.

(40) (C) Develop water-based or solvent-free processes.

Green chemistry is driving the development of water-based or solvent-free radical polymerization to reduce environmental impact. It ensures the sustainable production of next-generation polymers.

(41) (C) They hinder electrophilic substitution.

Friedel-Crafts alkylation does not occur in aromatic rings with strong electron-withdrawing groups, as EWGs reduce the electron density of the aromatic ring and make it less susceptible to electrophilic attack.

(42) (A) The introduction of an acyl group onto an aromatic ring.

Friedel-Crafts acylation is a specific type of EAS reaction where an acyl group (RCO–) is introduced onto an aromatic ring. It involves the use of an acyl halide (like acyl chloride, RCOCl).

(43) (B) Acylium ion (RCO^+).

Friedel-Crafts acylation introduces an acyl group (RCO–) onto the aromatic ring using an acyl chloride (RCOCl) or a carboxylic acid anhydride ($(RCO)_2O$) in the presence of a Lewis acid catalyst, usually aluminum chloride ($AlCl_3$). The electrophile is a resonance-stabilized acylium ion (RCO^+).

(44) (C) The acyl group is a deactivating group.

The introduction of an electron-withdrawing acyl group deactivates the aromatic ring toward further electrophilic attack and makes polyacylation less likely. Acyl groups are deactivating groups and reduce the reactivity of the acylated product.

(45) (D) It does not result in carbocation rearrangements.

In Friedel-Crafts acylation, acylium ions are resonance-stabilized and do not undergo rearrangements. This is a significant advantage over Friedel-Crafts alkylation.

(46) (A) Secondary amines.

Aldehydes and ketones react with secondary amines (R_2NH) to form enamines. This reaction also involves the nucleophilic addition of the amine to the carbonyl group, followed by the elimination of water.

(47) (D) Carbinolamine.

The mechanism of imine formation begins with the nucleophilic attack by a primary amine on the carbonyl carbon, which results in the formation of a tetrahedral intermediate known as a carbinolamine. Following this, a proton transfer occurs from the nitrogen atom to the oxygen atom. Finally, water is eliminated, which results in the generation of the imine.

(48) (C) Alcohol.

The alcohol functional group is characterized by the presence of a hydroxyl group (–OH) attached to a saturated carbon atom (sp^3 hybridized). This distinguishes it from other functional groups like ketones and esters, which contain carbonyl groups instead.

(49) (B) Iminium ion.

The formation of enamine starts with the nucleophilic attack by a secondary amine on the carbonyl carbon, which creates a tetrahedral intermediate called a carbinolamine. A proton is then transferred, which results in the elimination of water and yields an iminium ion. The iminium ion is vital in the next step, where deprotonation occurs to generate the enamine.

(50) (D) They can act as nucleophiles in various reactions.

Enamines are valuable in organic synthesis because they possess a carbon-carbon double bond adjacent to a nitrogen atom. This structure allows them to act as nucleophiles.

(51) (D) They can react at two different sites.

Enolates are ambident nucleophiles, which means they can react with electrophiles at two different sites: the α-carbon and the oxygen atom. The regioselectivity of enolate alkylation can be influenced by factors such as the nature of the electrophile, the solvent, and the counterion.

(52) (D) A new carbon-carbon bond.

In C-alkylation, the reaction at the α-carbon causes the formation of a new carbon-carbon bond. This is the most common and synthetically useful mode of reactivity. In O-alkylation, the reaction at the oxygen atom causes the formation of an enol ether. This is generally less desirable than C-alkylation.

(53) (A) β-hydroxyaldehydes or β-hydroxyketones.

Aldol reactions and Claisen condensations are two of the most important carbon-carbon bond-forming reactions in organic chemistry. They involve the reaction of an enolate with a carbonyl compound. The aldol reaction is the reaction of an enolate with an aldehyde or ketone to form a β-hydroxyaldehyde or β-hydroxyketone (i.e., an aldol).

(54) (D) Use only one carbonyl compound.

In a crossed aldol reaction, multiple products can form when two different carbonyl compounds are used. To minimize this and favor a single desired product, one effective strategy is to use only one carbonyl compound capable of forming the enolate ion while the other lacks α-hydrogens and thus cannot form an enolate, limiting the reaction pathway.

(55) (B) Two esters.

The Claisen condensation is the reaction of two esters to form a β-keto ester. This reaction is usually carried out using a strong base, such as sodium ethoxide (NaOEt). The mechanism of the Claisen condensation is similar to that of the aldol reaction.

(56) (A) The alkoxide intermediate is formed.

The mechanism of reduction with $LiAlH_4$ proceeds as follows.

1) Nucleophilic attack: The hydride ion (H^-) from $LiAlH_4$ attacks the electrophilic carbon atom of the carbonyl group, which forms an alkoxide intermediate.
2) Coordination: The aluminum atom coordinates to the oxygen atom of the alkoxide intermediate.
3) Further reduction: For carboxylic acids and esters, the reduction process involves multiple steps, with the initial reduction forming an aldehyde intermediate, which is then further reduced to an alcohol.
4) Protonation: After the reduction is complete, the reaction mixture is treated with dilute acid to protonate the alkoxide intermediate and give the alcohol product.

(57) (B) Aldehydes and ketones.

Sodium borohydride ($NaBH_4$) is often used in alcohol solvents such as methanol (MeOH) or ethanol (EtOH) or in water. It is a milder reducing agent than $LiAlH_4$ and can be used in protic solvents. It can reduce aldehydes and ketones to alcohols, but it does not usually reduce carboxylic acids, esters, or amides.

(58) (A) It is less expensive than $LiAlH_4$.

Sodium borohydride ($NaBH_4$) is a milder reducing agent compared to lithium aluminum hydride ($LiAlH_4$). It is more selective, which makes it useful for specific reactions. $NaBH_4$ can be used in protic solvents, such as alcohols or water, which adds to its versatility. Additionally, $NaBH_4$ is less expensive than $LiAlH_4$, which makes it a more economical choice for various reduction processes.

However, $NaBH_4$ has some limitations. It does not usually reduce carboxylic acids, esters, or amides. This restricts its use in specific reactions. Furthermore, the reaction rate of $NaBH_4$ is slower than that of $LiAlH_4$. This can affect the efficiency of the reduction process in some cases.

(59) (D) The reaction mixture is treated with dilute acid to protonate the alkoxide.

The mechanism of reduction with $NaBH_4$ is similar to that of $LiAlH_4$. It involves the nucleophilic attack of a hydride ion on the carbonyl carbon.

1) Nucleophilic attack: The hydride ion (H^-) from $NaBH_4$ attacks the electrophilic carbon atom of the carbonyl group, which forms an alkoxide intermediate.
2) Coordination: The boron atom coordinates to the oxygen atom of the alkoxide intermediate.
3) Protonation: After the reduction is complete, the reaction mixture is treated with dilute acid to protonate the alkoxide intermediate and give the alcohol product.

(60) (D) Sodium (Na).

Catalytic hydrogenation is the addition of hydrogen (H_2) to a molecule in the presence of a metal catalyst. This reaction is commonly used to reduce alkenes, alkynes, and carbonyl compounds. Common catalysts include palladium (Pd), platinum (Pt), nickel (Ni), and rhodium (Rh). The catalyst is usually supported on a high-surface-area material such as carbon (C). Commonly used solvents are alcohol solvents such as ethanol (EtOH), methanol (MeOH), or ethyl acetate (EtOAc).

(61) (B) Ensure it can be attached without extensive procedures.

Protecting groups are temporary modifications to a functional group that prevent it from undergoing unwanted reactions during chemical synthesis. The ideal protecting group should be easily attached to the functional group.

(62) (B) It remains intact during the synthesis.

A protecting group is used in organic synthesis to mask a reactive functional group temporarily. It must be stable under the reaction conditions employed during the synthesis for it to be effective.

(63) (D) Avoids affecting other functional groups during deprotection.

The protecting group should be easily removable under mild conditions without affecting other functional groups in the molecule. Additionally, the protecting group should be relatively inexpensive.

(64) (A) Introduce new stereocenters in a molecule.

Functional group interconversions perform the following functions.

- Modify reactivity: FGIs can introduce or remove functional groups to modify a molecule's reactivity.
- Protect reactive sites: FGIs can be used to protect reactive functional groups that would interfere with the desired transformations.

- Introduce stereocenters: Some FGIs can be used to introduce stereocenters into a molecule.
- Simplify synthesis: FGIs can convert complex functional groups into more basic ones, which simplifies a synthesis.

(65) (D) Prevent unwanted reactions of the alcohol.

Silyl ethers (TMS, TBS, TBDPS) are introduced by a reaction with a silyl chloride (e.g., TMSCl, TBSCl) in the presence of a base. They are removed by treatment with fluoride ions (e.g., TBAF, HF). Silyl ethers are stable to a wide range of reaction conditions, which include acids, bases, and oxidizing agents.

(66) (C) Smart packaging.

The packaging sector is embracing innovation. Biodegradable polymers, such as polylactic acid (PLA) and polyhydroxyalkanoates (PHA), are gaining popularity as sustainable alternatives, with PLA being used in compostable coffee cups. Smart packaging, which incorporates polymer-based sensors or time-temperature indicators, monitors the freshness of food. Meanwhile, multilayer films that combine polyethylene (PE) with ethylene vinyl alcohol (EVOH) improve vacuum-sealed packaging. However, plastic waste remains a significant challenge. This issue has prompted advancements in recyclable polymers and chemical recycling, which breaks polymers into reusable monomers.

(67) (C) PVC.

Polymers play a critical role in construction, with PVC being a cornerstone for pipes, window frames, siding, and roofing membranes. PVC's durability, resistance to weathering, and low cost make it ideal for long-term use in harsh environments.

(68) (C) A chain of repeated styrene units with pendant phenyl groups.

Polystyrene is formed by polymerizing styrene monomers, which results in a carbon backbone chain with phenyl (benzene) rings attached as side groups (pendant groups). This gives polystyrene its rigidity and clarity, commonly seen in products like disposable cutlery and packaging materials.

(69) (D) Bridges and structural components.

Polymers, such as FRPs, are used in construction. Its applications have resulted in numerous advancements in engineering functionality, safety, and economy, chiefly due to its remarkable mechanical properties. For example, carbon-fiber-reinforced epoxy is used in the construction of bridges and structural components.

(70) (C) Polypropylene.

In the packaging industry, polymers such as polyethylene (PE), polypropylene (PP), and polystyrene (PS) are dominant due to their low cost and adaptability. PP is used to form rigid containers like yogurt cups. It provides barrier properties against moisture, gases, and UV light, which extends the shelf life of food, beverages, and other products. Its ability to be molded into various shapes and compatibility with printing for branding further enhance its utility.

Test 5: Questions

(1) What is the proper IUPAC name for a carboxylic acid that contains four carbon atoms?

(A) Butanal.

(B) Butanone.

(C) Butanoic acid.

(D) Butanamide.

(2) Which prefix is designated for a nitrile group when it is not the principal functional group of a compound?

(A) Cyano-.

(B) Formyl-.

(C) Carboxy-.

(D) Carbamoyl-.

(3) How should a compound that contains two alcohol groups be named according to IUPAC conventions?

(A) With the suffix -ol.

(B) With the suffix -diol.

(C) With the prefix hydroxy-.

(D) With the prefix dihydroxy-.

(4) According to IUPAC naming rules, which of the following correctly describes the compound bicyclo[2.2.1]heptane?

(A) A seven-carbon bicyclic compound.

(B) A five-carbon bicyclic compound.

(C) A nine-carbon bicyclic compound.

(D) A six-carbon bicyclic compound.

(5) How is the compound 1,3-dichlorobenzene referred to within the ortho/meta/para nomenclature system?

(A) Ortho-dichlorobenzene.

(B) Meta-dichlorobenzene.

(C) Para-dichlorobenzene.

(D) Mono-dichlorobenzene.

(6) How does a π bond affect the reactivity of a molecule?

(A) Reduces the overall reactivity of the molecule.

(B) Prevents substitution reactions from occurring.

(C) Enables addition reactions due to its electron-rich nature.

(D) Stabilizes the molecule completely and negates any reactivity.

(7) Which spectroscopic property is notably influenced by the presence of conjugation in a molecule?

(A) Infrared (IR) absorption only.

(B) Ultraviolet-visible (UV-Vis) absorption.

(C) NMR chemical shifts only.

(D) Mass spectrometry peaks.

(8) What is a common mistake made when identifying the hybridization state of a carbon atom?

(A) Check bond angles to assess hybridization accurately.

(B) Assume that all carbon atoms are sp^3 hybridized.

(C) Count the number of electron pairs around the atom.

(D) Consider the presence of lone pairs when determining hybridization.

(9) Which molecule exhibits a linear geometry due to sp hybridization?

(A) But-2-ene.

(B) But-1-yne.

(C) Butane.

(D) Acetone.

(10) How does resonance influence the bond lengths within the carbonate ion (CO_3^{2-})?

(A) It creates unequal bond lengths between the carbon and oxygen atoms.

(B) It results in identical bond lengths for all C–O bonds in the ion.

(C) It eliminates the presence of double bonds entirely in the structure.

(D) It increases the overall bond strength of the carbon-oxygen bonds.

(11) How do Brønsted and Lewis acids act as catalysts in chemical reactions?

(A) They generate a stronger nucleophile.

(B) They deprotonate a reactant.

(C) They protonate a reactant.

(D) They activate a leaving group.

(12) How does a higher s-character in the hybrid orbital affect the acidity of C–H bonds?

(A) Decreases acidity.

(B) Increases acidity.

(C) Does not affect acidity.

(D) Increases basicity.

(13) Why are meso compounds considered achiral despite having chiral centers?

(A) Due to their high reactivity.

(B) Because they contain isotopes.

(C) Because they possess internal symmetry elements.

(D) Because they are optically active.

(14) What effect does the internal symmetry in meso compounds have on their optical activity?

(A) Enhances optical activity.

(B) Cancels out optical activity.

(C) Increases optical rotation.

(D) Induces optical isomerism.

(15) Which of the following chiral-center configurations is necessary for a meso compound?

(A) Both centers must be (R).

(B) Both centers must be (S).

(C) One center must be (R), and the other must be (S).

(D) The centers can have any configuration.

(16) What is the maximum number of stereoisomers in a molecule with two chiral centers?

(A) Two.

(B) Three.

(C) Four.

(D) Six.

(17) How do meso compounds differ from their chiral counterparts in terms of optical activity?

(A) Meso compounds are optically active.

(B) Meso compounds are optically inactive.

(C) Both are optically inactive.

(D) Both are optically active.

(18) What effect does symmetry have on the number of stereoisomers?

(A) Increases the number of stereoisomers.

(B) Decreases the number of stereoisomers.

(C) Has no effect on stereoisomers.

(D) Only affects meso compounds.

(19) Which of the following statements about odd numbers of chiral centers in meso compounds is true?

(A) They are always meso compounds.

(B) They rarely result in meso compounds.

(C) They never form meso compounds.

(D) Odd numbers facilitate the formation of a plane of symmetry.

(20) What does an even number of stereocenters in meso compounds help accomplish?

(A) Facilitate the formation of a plane of symmetry.

(B) Ensure they are always optically active.

(C) Decrease the number of possible stereoisomers.

(D) Ensure they have no chiral centers.

(21) What type of nucleophiles promotes SN2 reactions?

(A) Weak nucleophiles.

(B) Strong nucleophiles.

(C) Sterically hindered nucleophiles.

(D) Neutral nucleophiles.

(22) Which of the following is the correct order of halides as leaving groups from best to worst?

(A) $Cl^- > Br^- > I^- > F^-$.

(B) $I^- > Br^- > Cl^- > F^-$.

(C) $F^- > Cl^- > Br^- > I^-$.

(D) $Br^- > I^- > Cl^- > F^-$.

(23) What type of solvent do SN2 and E2 reactions generally favor?

(A) Polar protic.

(B) Non-polar.

(C) Polar aprotic.

(D) Aqueous.

(24) Why are strongly basic leaving groups generally poor leaving groups?

(A) They can easily depart as stable species.

(B) They are rarely displaced under usual conditions.

(C) They stabilize cations.

(D) They increase reaction rates.

(25) What effect do lower temperatures generally have on substitution and elimination reactions?

(A) Favor substitution.

(B) Favor elimination.

(C) Have no effect.

(D) Increase the reaction rate for both.

(26) What does stereoselectivity refer to in a chemical reaction?

(A) Preference for forming one stereoisomer over others.

(B) Preference for forming constitutional isomers.

(C) Preference for forming alkanes.

(D) Temperature dependence.

(27) Which type of addition occurs when two groups add to the same face of an alkene?

(A) Anti addition.

(B) Syn addition.

(C) Pseudo addition.

(D) Substitution.

(28) Which of the following reactions is an example of syn addition?

(A) Bromination of alkenes.

(B) SN1 reaction.

(C) Hydroboration.

(D) Dehydrohalogenation.

(29) What does it mean for a reaction to be stereospecific?

(A) The reaction produces only one product.

(B) The stereochemistry of the reactant determines the stereochemistry of the product.

(C) The reaction is temperature-dependent.

(D) The reaction does not involve stereoisomers.

(30) Which of the following statements referring to anti addition in the context of alkenes is correct?

(A) Groups add to the same face.

(B) No addition occurs.

(C) Only one group adds.

(D) Groups add to opposite faces.

(31) Why is ^{13}C NMR considered to have lower sensitivity compared to ^{1}H NMR?

(A) ^{13}C has a higher magnetic moment.

(B) ^{13}C is less abundant and has a smaller gyromagnetic ratio.

(C) ^{13}C nuclei are more stable.

(D) ^{13}C does not couple with protons.

(32) What technique can be used to determine the number of hydrogens attached to each carbon in ^{13}C NMR?

(A) Distortionless enhancement by polarization transfer (DEPT).

(B) APT.

(C) COSY.

(D) HSQC.

(33) Which of the following statements regarding the effect of broadband decoupling in ^{13}C NMR is correct?

(A) It increases the number of signals.

(B) It simplifies the spectrum by removing coupling.

(C) It enhances the signal intensity.

(D) It causes overlapping of signals.

(34) Which of the following statements regarding the benefit of using ^{13}C NMR over ^{1}H NMR is correct?

(A) Higher sensitivity to small changes.

(B) More peaks in the spectrum.

(C) Easier identification of hydrogen atoms.

(D) Direct information about the carbon skeleton.

(35) What happens to the signals in proton-decoupled ^{13}C NMR spectra?

(A) They appear as doublets.

(B) They appear as triplets.

(C) They appear as singlets.

(D) They disappear entirely.

(36) What is the primary advantage of polymers synthesized through radical polymerization?

(A) They are limited to forming rigid plastics only.

(B) They can only be used for low-temperature applications.

(C) They form highly crystalline structures unsuitable for flexibility.

(D) They allow for tailored properties to suit a wide range of applications.

(37) What was the main use of chlorofluorocarbons (CFCs) before being phased out?

(A) Fuels.

(B) Solvents.

(C) Pesticides.

(D) Refrigerants.

(38) Why are non-halogenated solvents increasingly preferred in green organic chemistry practices?

(A) They react readily with most functional groups.

(B) They enhance the reactivity of halogenated compounds.

(C) They are less toxic and more environmentally friendly than halogenated solvents.

(D) They increase the yield of all organic reactions.

(39) What is a significant drawback of radical halogenation in industrial applications?

(A) It often lacks selectivity.

(B) It produces only one product.

(C) It requires rare materials.

(D) It is too complex.

(40) How do advances in photocatalysis impact radical halogenation?

(A) They decrease energy efficiency.

(B) They eliminate the need for light.

(C) They improve selectivity and energy efficiency.

(D) They increase the number of byproducts.

(41) What is the main function of activating groups in electrophilic aromatic substitution (EAS)?

(A) Decrease electron density.

(B) Stabilize carbocations.

(C) Increase the electron density of the aromatic ring.

(D) Deactivate the ring toward electrophiles.

(42) What is the main mechanism through which strongly activating groups donate electrons?

(A) Inductive effects.

(B) Hyperconjugation.

(C) Resonance effects.

(D) Hydrogen bonding.

(43) What distinguishes weakly activating groups from strongly activating groups in aromatic substitution reactions?

(A) Weakly activating groups mainly use hyperconjugation and inductive effects.

(B) Weakly activating groups do not donate electrons.

(C) Weakly activating groups are only found in alkyl chains.

(D) Weakly activating groups are always deactivating.

(44) Which of the following would be classified as a moderately activating group in aromatic substitution reactions?

(A) Hydroxyl group (–OH).

(B) Alkoxy groups (–OR).

(C) Amides (–NHCOR).

(D) Amino group ($-NH_2$).

(45) What effect do deactivating groups have on the electron density of the aromatic ring?

(A) They increase electron density.

(B) They have no effect on electron density.

(C) They decrease electron density.

(D) They stabilize the ring.

(46) Which type of compounds are transformed by the Wittig reaction?

(A) Alkenes into aldehydes.

(B) Aldehydes and ketones into alkenes.

(C) Alkenes into alcohols.

(D) Alcohols into carbonyl compounds.

(47) Which of the following compounds is a key component of the Wittig reaction?

(A) Phosphorus ylide.

(B) Grignard reagent.

(C) Triphenylphosphine oxide.

(D) Carbon dioxide.

(48) Which of the following is the general formula for Grignard reagents?

(A) RMgX.

(B) R2C=O.

(C) RCHO.

(D) RCOOH.

(49) Which property mainly contributes to the high boiling points of carboxylic acids?

(A) Ionic bonding.

(B) Hydrogen bonding.

(C) Van der Waals forces.

(D) Covalent bonding.

(50) All of the following are derivatives of carboxylic acids except:

(A) Esters.

(B) Amides.

(C) Alkanes.

(D) Acid halides.

(51) What drives the equilibrium toward product formation in the Claisen condensation?

(A) Deprotonation of the β-keto ester by base.

(B) Formation of a stable cyclic product.

(C) Increase in temperature.

(D) Removal of solvents.

(52) What is a key characteristic of the Claisen condensation regarding its reversibility?

(A) It is always irreversible.

(B) The equilibrium usually lies toward the starting materials.

(C) It cannot reach equilibrium.

(D) It only occurs at high temperatures.

(53) What is the main product of the Dieckmann condensation?

(A) β-hydroxy aldehyde.

(B) β-keto ester.

(C) Dicarbonyl compound.

(D) Simple ester.

(54) Why is the Dieckmann condensation regarded as advantageous in synthetic chemistry?

(A) It produces simple alcohols.

(B) It allows for the formation of linear chains.

(C) It forms cyclic β-keto esters that are precursors for biologically active molecules.

(D) It requires no catalysts.

(55) What is the role of the α,β-unsaturated carbonyl compound in a Michael addition?

(A) It acts as an electrophile.

(B) It acts as a nucleophile.

(C) It acts as a solvent.

(D) It acts as a catalyst.

(56) What happens during the activation step of catalytic hydrogenation?

(A) The organic molecule is reduced.

(B) The H–H bond of the hydrogen molecule is broken.

(C) The catalyst is deactivated.

(D) The solvent evaporates.

(57) Which statement regarding the limitation of catalytic hydrogenation is correct?

(A) It can only reduce alkenes.

(B) It does not require a catalyst.

(C) It is always non-selective.

(D) It requires high pressure of hydrogen gas.

(58) How does PCC compare to CrO_3 in terms of chemoselectivity?

(A) PCC is less selective than CrO_3.

(B) PCC is more selective.

(C) Both are equally selective.

(D) CrO_3 is more selective than PCC.

(59) How can catalytic hydrogenation be made stereoselective?

(A) Utilize a non-chiral catalyst.

(B) Increase the temperature.

(C) Utilize a higher pressure of hydrogen.

(D) Utilize a chiral catalyst or chiral auxiliary.

(60) What is the main function of Corey-House synthesis?

(A) Oxidize alcohols.

(B) Construct alkenes from alkanes.

(C) Couple alkyl halides to form alkanes.

(D) Introduce chirality into molecules.

(61) Which of the following is the correct first step in employing protecting groups?

(A) Introduce protecting groups.

(B) Identify functional groups that may interfere with transformations.

(C) Remove existing protecting groups.

(D) Analyze the final product.

(62) All of the following factors should be considered when selecting a protecting group except:

(A) Ease of introduction.

(B) Stability under reaction conditions.

(C) Molecular weight.

(D) Ease of removal.

(63) Which of the following statements regarding the advantage of introducing protecting groups early in a synthesis is correct?

(A) It minimizes the time reactive groups are unprotected.

(B) It complicates the reaction conditions.

(C) It increases the overall reaction time.

(D) It allows for more complex reactions.

(64) What is the recommendation regarding the timing of the removal of protecting groups in the synthetic sequence?

(A) As early as possible.

(B) At the midpoint of the synthesis.

(C) As late as possible.

(D) Whenever convenient.

(65) Which of the following is a commonly used protecting group for amines?

(A) Boc (tert-butyloxycarbonyl).

(B) TMS (trimethylsilyl).

(C) Acetal.

(D) Ester.

(66) Which of the following synthetic polymers is commonly used in hosiery and ropes?

(A) Polyester.

(B) Acrylic.

(C) Nylon.

(D) Polypropylene.

(67) Which of the following is Kevlar mainly used for in the textile industry?

(A) Apparel.

(B) Bulletproof vests.

(C) Upholstery.

(D) Blankets.

(68) What property of certain polymers makes them ideal for manufacturing processes involving shaping and forming?

(A) High crystallinity.

(B) Ease of molding.

(C) High thermal conductivity.

(D) Electrical insulation.

(69) Which polymer is known for its thermal stability in battery housings for electric vehicles?

(A) Polyamide.

(B) Polypropylene.

(C) Nylon.

(D) Polyurethane.

(70) Which polymer is known for its use in 3D-printed implants for bone regeneration?

(A) Polyethylene.

(B) Polylactic-co-glycolic acid (PLGA).

(C) Polycaprolactone (PCL).

(D) Polyimide.

Test 5: Answers and Explanations

(1) (C) Butanoic acid.

To name carboxylic acids, chemists identify the longest carbon chain that contains the carboxyl group (–COOH) and add the suffix -oic. For a carboxylic acid with four carbon atoms, the parent name is derived from butane, which results in butanoic acid.

(2) (A) Cyano-.

When a nitrile group (–C≡N) is not the principal functional group in a compound, it is designated with the prefix cyano-. For example, in a compound like 2-cyanoethanol, the cyano group is treated as a substituent, and cyano- indicates its presence.

(3) (B) With the suffix -diol.

When a compound contains two alcohol (–OH) groups, the IUPAC naming convention uses the suffix -diol to indicate the presence of two hydroxyl groups. For a compound like butanediol, the -diol suffix signifies that there are two hydroxyl groups in the butane structure.

(4) (A) A seven-carbon bicyclic compound.

The IUPAC name indicates that the compound is bicyclic (having two fused rings). The notation [2.2.1] describes the number of carbon atoms in each segment of the bicyclic structure. In this case, the compound contains a total of seven carbon atoms and will be named as a derivative of heptane.

(5) (B) Meta-dichlorobenzene.

In this compound, the two chlorine substituents are located on the benzene ring at the 1 and 3 positions. The terms ortho, meta, and para refer to the relative positions of substituents on a benzene ring. In this case, the chlorines are separated by one carbon atom, which corresponds to the meta designation.

(6) (C) Enables addition reactions due to its electron-rich nature.

A π bond, particularly in alkenes, is reactive and facilitates addition reactions because it possesses higher electron density. This bond increases reactivity rather than decreasing it. Substitution reactions can occur in aromatic compounds. Although resonance can stabilize a structure, it does not eliminate reactivity.

(7) (B) Ultraviolet-visible (UV-Vis) absorption.

The presence of conjugation in a molecule, which involves alternating double and single bonds, results in a system of π electrons that can be delocalized. This delocalization lowers the energy gap between the ground and excited states of the electrons. This results in the absorption of longer wavelengths of light in the UV-Vis region. Conjugated systems absorb light at specific wavelengths, which can be analyzed to provide information about the structure of the molecule.

(8) (B) Assume that all carbon atoms are sp^3 hybridized.

A frequent error in hybridization analysis is the assumption that all carbon atoms are sp^3 hybridized, which overlooks the fact that carbon atoms involved in double or triple bonds will exhibit sp^2 or sp hybridization, respectively.

(9) (B) But-1-yne.

But-1-yne (C_4H_6) contains an sp-hybridized carbon atom due to the presence of a triple bond, which results in a linear molecular geometry. But-2-ene is a molecule with sp^2 hybridization, and butane consists of sp^3-hybridized carbons. Acetone features sp^2-hybridized carbon atoms, which makes it non-linear.

(10) (B) It results in identical bond lengths for all C–O bonds in the ion.

Resonance in the carbonate ion causes the delocalization of electrons. This reaction makes all C–O bonds equivalent and results in identical bond lengths. Resonance mainly affects bond lengths rather than just bond strength.

(11) (C) They protonate a reactant.

Both Brønsted and Lewis acids can act as catalysts by protonating a reactant to enhance its electrophilicity (making it more susceptible to nucleophilic attack) or by stabilizing a leaving group. For example, in the acid-catalyzed hydration of an alkene, the acid (H_3O^+) protonates the alkene, which forms a more electrophilic carbocation intermediate that is subsequently attacked by water.

(12) (B) Increases acidity.

The hybridization of the atom holding the lone pair in the conjugate base (or the atom losing the proton) affects acidity. Higher s-character in the hybrid orbital means the electrons are held closer to the nucleus, which stabilizes the conjugate base. Therefore, the acidity of C–H bonds increases in the order $sp^3 < sp^2 < sp$. For example, terminal alkynes are more acidic than alkenes, which are more acidic than alkanes.

(13) (C) Because they possess internal symmetry elements.

Meso compounds are a unique class of achiral molecules despite containing chiral centers (stereocenters). This apparent contradiction arises because meso compounds possess an internal plane of symmetry or center of symmetry, which makes them superimposable on their mirror images.

(14) (B) Cancels out optical activity.

Due to their symmetry, meso compounds do not rotate plane-polarized light, as the rotations from the chiral centers cancel out. This distinguishes them from chiral stereoisomers, which are optically active.

(15) (C) One center must be (R), and the other must be (S).

The chiral centers must be structurally identical in terms of their substituents (i.e., both carbons must be bonded to the same four groups in a symmetric arrangement). The configurations at these centers must be opposite (i.e., one R and one S), so their chiral contributions cancel out due to the molecule's symmetry.

(16) (C) Four.

For molecules with n chiral centers, the maximum number of stereoisomers is 2^n. For example, a molecule with two chiral centers could have up to four stereoisomers (two enantiomeric pairs). However, meso compounds reduce this number by introducing symmetry. If a meso form exists, there are only three stereoisomers (one meso and one enantiomeric pair).

(17) (B) Meso compounds are optically inactive.

Meso compounds are optically inactive due to the cancellation of rotation from opposite-configured chiral centers. Conversely, chiral compounds are optically active as they rotate plane-polarized light. For example, (R,R)-tartaric acid is dextrorotatory, and (S,S)-tartaric acid is levorotatory.

(18) (B) Decreases the number of stereoisomers.

Meso compounds reduce the number of stereoisomers by introducing symmetry. For example, a molecule with two chiral centers could have up to four stereoisomers (two enantiomeric pairs). However, if a meso form exists, there are only three stereoisomers (one meso, one enantiomeric pair).

(19) (B) They rarely result in meso compounds.

Meso compounds often have an even number of chiral centers (e.g., two or four), as this facilitates the formation of a plane of symmetry. However, in rare cases, odd numbers of chiral centers can result in meso compounds.

(20) (A) Facilitate the formation of a plane of symmetry.

When a compound has an even number of stereocenters (usually two), it can be arranged such that one stereocenter's configuration is the opposite of the other (e.g., one R and one S). This arrangement allows for the possibility of a plane of symmetry.

(21) (B) Strong nucleophiles.

Strong nucleophiles promote SN2 reactions, while strong bases favor E2 reactions. The basicity and nucleophilicity of a reagent are related but not identical. For example, bulky strong bases like tert-butoxide are poor nucleophiles and tend to favor elimination (E2) over substitution.

(22) (B) $I^- > Br^- > Cl^- > F^-$.

A good leaving group is critical for both substitution and elimination reactions. It can depart as a stable, weakly basic species that carries the electron pair from the cleaved bond. Common good leaving groups include the following.

- Halides: $I^- > Br^- > Cl^- > F^-$ (the order may vary in polar protic solvents due to solvation effects).
- Sulfonates: Tosylate (OTs^-), mesylate (OMs^-), and triflate (OTf^-) are excellent leaving groups.
- Water (H_2O) or protonated alcohols: These can serve as good leaving groups when the oxygen is protonated.
- Nitrogen gas (N_2): An excellent leaving group in diazonium salt reactions.

(23) (C) Polar aprotic.

Polar aprotic solvents, like acetone, DMSO, and DMF, have a significant dipole moment but lack acidic hydrogen atoms. They can effectively solvate cations while having weaker interactions with anions. They generally favor SN2 and E2 reactions by keeping the anionic nucleophile/base reactive.

(24) (B) They are rarely displaced under usual conditions.

Strongly basic leaving groups (e.g., OH^-, RO^-, NH_2^-) are generally poor leaving groups and are rarely displaced in substitution or elimination reactions under the usual conditions.

(25) (A) Favor substitution.

Lower temperatures generally favor substitution over elimination due to reduced activation energy. Conversely, higher temperatures favor elimination reactions due to a more favorable entropy change ($\Delta S^\circ > 0$ for elimination, as one molecule forms two or more).

(26) (A) Preference for forming one stereoisomer over others.

Stereoselectivity refers to the preference for forming one stereoisomer over others when multiple stereoisomers can be produced. This preference can significantly influence the properties and reactivity of the resulting compounds.

(27) (B) Syn addition.

Syn addition occurs when two atoms or groups add to the same face of the alkene or alkyne. Hydroboration is a syn addition, as is catalytic hydrogenation (addition of H_2 with a metal catalyst).

(28) (C) Hydroboration.

In hydroboration, boron and hydrogen add across the double bond of an alkene. This addition occurs in a syn manner. This means that both the boron and hydrogen add to the same face of the alkene.

(29) (B) The stereochemistry of the reactant determines the stereochemistry of the product.

A reaction is stereospecific if the stereochemistry of the reactant dictates the stereochemistry of the product. For example, the SN2 reaction is stereospecific, as a chiral substrate will always undergo inversion of configuration.

(30) (D) Groups add to opposite faces.

Anti addition occurs when two atoms or groups add to opposite faces of the alkene or alkyne. Bromination of alkenes follows an anti-addition mechanism that involves a cyclic bromonium ion intermediate.

(31) (B) ^{13}C is less abundant and has a smaller gyromagnetic ratio.

^{13}C NMR is less sensitive than ^{1}H NMR due to a combination of factors: the lower natural abundance of ^{13}C (1.1%) compared to ^{1}H (almost 100%) and a smaller gyromagnetic ratio for ^{13}C (one-quarter of that of ^{1}H).

(32) (A) Distortionless enhancement by polarization transfer (DEPT).

In simple ^{13}C NMR spectra (proton-decoupled), signals appear as singlets because the coupling between ^{13}C nuclei and protons is removed through broadband decoupling. This simplifies the spectrum, which makes it easier to count non-equivalent carbon atoms. Techniques like DEPT can be employed to determine the number of hydrogens attached to each carbon.

(33) (B) It simplifies the spectrum by removing coupling.

Broadband decoupling is a technique used in ^{13}C NMR to eliminate the coupling between ^{13}C nuclei and protons. This results in the carbon signals appearing as singlets rather than multiplets.

(34) (D) Direct information about the carbon skeleton.

^{13}C is a less abundant isotope of carbon (about 1.1%) and has lower sensitivity in NMR compared to ^{1}H. However, ^{13}C NMR provides direct information about a molecule's carbon skeleton.

(35) (C) They appear as singlets.

In proton-decoupled ^{13}C NMR, the coupling between ^{13}C nuclei and surrounding protons is removed through a technique called broadband decoupling. This means that each unique carbon atom generates a single signal (singlet) rather than multiple peaks (multiplets) that would result from coupling interactions.

(36) (D) They allow for tailored properties to suit a wide range of applications.

Radical polymerization enables the creation of polymers with customized features such as flexibility, strength, thermal resistance, and chemical reactivity. This versatility allows these polymers to be adapted for diverse uses, including packaging, biomedical devices, insulation, and more.

(37) (D) Refrigerants.

Radical halogenation of alkanes is usually initiated by light or heat. It introduces halogen atoms (e.g., chlorine or bromine) into hydrocarbons, which produce halogenated compounds like chloroform or chloromethane. These serve as intermediates for solvents, refrigerants, and agrochemicals. For example, CFCs, once widely used as refrigerants, were synthesized through radical chlorination, though their environmental impact led to their discontinuation.

(38) (C) They are less toxic and more environmentally friendly than halogenated solvents.

Non-halogenated solvents are favored in sustainable chemistry because they pose fewer environmental and health risks compared to halogenated solvents, which are often persistent, bioaccumulative, and potentially carcinogenic.

(39) (A) It often lacks selectivity.

Radical halogenation of alkanes is valued for its simplicity and ability to functionalize inert alkanes. However, it often lacks selectivity and produces mixtures of products that require purification.

(40) (C) They improve selectivity and energy efficiency.

Photocatalysis utilizes light to activate radical initiators, which can result in more controlled and selective reactions. This means that instead of producing a mixture of products, the process can be fine-tuned to favor the formation of desired halogenated compounds.

(41) (C) Increase electron density of the aromatic ring.

Activating groups increase the electron density of the aromatic ring, which makes it more nucleophilic and thus more reactive toward electrophiles compared to benzene. They achieve this through electron donation through resonance (+M or +R effect) or inductive effects (+I effect).

(42) (C) Resonance effects.

These groups possess lone pairs of electrons directly attached to the ring that can be readily donated through resonance. This significantly increases electron density, especially at the ortho and para positions.

(43) (A) Weakly activating groups mainly use hyperconjugation and inductive effects.

These groups mainly donate electrons through hyperconjugation (overlap of σ bonds with the π system) and inductive effects (+I) due to the electron-releasing nature of alkyl groups.

(44) (C) Amides (–NHCOR).

Moderately activating groups donate electrons through resonance, but the lone pairs are involved in resonance elsewhere, which makes their donation to the ring less effective. Examples include amides (–NHCOR) and esters (–OCOR).

(45) (C) They decrease electron density.

Deactivating groups decrease the electron density of the aromatic ring, which makes it less nucleophilic and thus less reactive toward electrophiles compared to benzene. They achieve this through electron withdrawal through resonance (–M or –R effect) or inductive effects (–I effect).

(46) (B) Aldehydes and ketones into alkenes.

The Wittig reaction is a powerful method for converting aldehydes and ketones into alkenes. In this reaction, a carbonyl compound reacts with a phosphorus ylide (i.e., a Wittig reagent).

(47) (A) Phosphorus ylide.

A phosphorus ylide is a compound containing a negatively charged carbon atom bonded to a positively charged phosphorus atom. Wittig reagents are usually prepared by reacting a triphenylphosphonium salt with a strong base.

(48) (A) RMgX.

Grignard reagents (RMgX, where X is a halogen) are organometallic compounds that are strong nucleophiles and strong bases. They react with aldehydes and ketones to form alcohols.

(49) (B) Hydrogen bonding.

Carboxylic acids can form strong hydrogen bonds with themselves and with other molecules. This results in relatively high boiling points and good solubility in polar solvents.

(50) (C) Alkanes.

Carboxylic acid derivatives are compounds in which the hydroxyl group of the carboxyl group has been replaced by another group. Common carboxylic acid derivatives include esters (RCOOR'), amides (RCONR'R"), acid halides (RCOX, where X = Cl, Br), and anhydrides (RCOOCOR').

(51) (A) Deprotonation of the β-keto ester by base.

The β-keto ester is deprotonated by the base, which forms a stable enolate anion. This deprotonation is essential for driving the equilibrium toward product formation.

(52) (B) The equilibrium usually lies toward the starting materials.

The Claisen condensation is reversible, and the equilibrium usually lies toward the starting materials. However, if the β-keto ester product has an acidic α-proton, it can be deprotonated by the base, which forms a stable enolate anion. This deprotonation shifts the equilibrium toward product formation.

(53) (B) β-keto ester.

The Dieckmann condensation is an intramolecular Claisen condensation. This reaction is particularly useful for forming cyclic β-keto esters. In the Dieckmann condensation, a diester undergoes deprotonation at one of its α-carbons by a strong base, which generates an enolate ion. This enolate then attacks the carbonyl carbon of another ester group within the same molecule, which results in the formation of a cyclic intermediate. Upon hydrolysis, this intermediate yields a β-keto ester, which is characterized by the presence of both a ketone and an ester functional group within a cyclic structure.

(54) (C) It forms cyclic β-keto esters that are precursors for biologically active molecules.

The ability to form cyclic β-keto esters through the Dieckmann condensation is particularly advantageous in synthetic chemistry. These compounds can serve as precursors for various biologically active molecules and are also useful in the synthesis of complex natural products.

(55) (A) It acts as an electrophile.

The α,β-unsaturated carbonyl compound acts as an electrophile, with the β-carbon being the electrophilic site. The nucleophile can be a variety of species, which include enolates, amines, thiols, and cyanide.

(56) (B) The H–H bond of the hydrogen molecule is broken.

The mechanism of catalytic hydrogenation proceeds as follows.

1) Adsorption: The hydrogen molecule and the organic molecule adsorb onto the surface of the metal catalyst.
2) Activation: The hydrogen molecule is activated by the metal catalyst, which breaks the H–H bond and forms metal-hydrogen bonds.
3) Transfer: The hydrogen atoms are transferred to the organic molecule, which reduces the double or triple bond.
4) Desorption: The reduced product desorbs from the surface of the metal catalyst.

(57) (D) It requires high pressure of hydrogen gas.

The following are the limitations of catalytic hydrogenation.

- It often requires high pressure of hydrogen gas.
- It can reduce multiple functional groups, which can be a limitation if selectivity is desired.
- The stereochemistry of the product can be difficult to control.

(58) (B) PCC is more selective.

Chemoselectivity refers to the ability of a reagent to react preferentially with one functional group over another. PCC is chemoselective for oxidizing alcohols to aldehydes or ketones without oxidizing other functional groups, while CrO_3 is less selective. $NaBH_4$ is chemoselective for reducing aldehydes and ketones without reducing carboxylic acids, esters, or amides, while $LiAlH_4$ is less selective. Additionally, by carefully choosing the catalyst and reaction conditions, catalytic hydrogenation can be made chemoselective for reducing certain functional groups.

(59) (D) Utilize a chiral catalyst or chiral auxiliary.

Stereoselectivity refers to the ability of a reagent to form one stereoisomer preferentially over another. Catalytic hydrogenation can be stereoselective if the catalyst is chiral or if the reaction is carried out in the presence of a chiral auxiliary. The reduction of a cyclic ketone can be stereoselective, with the hydride ion attacking preferentially from one face of the ring.

(60) (C) Couple alkyl halides to form alkanes.

Oxidation and reduction reactions are often used in multi-step syntheses to construct complex molecules. The Corey-House synthesis utilizes organocopper reagents to couple alkyl halides, which then form alkanes.

(61) (B) Identify functional groups that may interfere with transformations.

The first step in employing protecting groups is to identify any functional groups within the molecule that may interfere with the desired transformations. Reactive functional groups (such as alcohols, amines, and carboxylic acids) can participate in side reactions that complicate the synthetic process.

(62) (C) Molecular weight.

Once reactive functional groups have been identified, the next step is to select appropriate protecting groups that are compatible with the reaction conditions used in the synthesis. The choice of protecting group depends on several factors. These include the stability of the group under various reaction conditions, its ease of introduction and removal, and its influence on the overall reactivity of the molecule.

(63) (A) It minimizes the time reactive groups are unprotected.

It is advisable to introduce protecting groups as early as possible in the synthesis to maximize efficiency. By doing so, chemists can minimize the duration for which reactive functional groups remain unprotected, which reduces the risk of unwanted side reactions. The early introduction of protecting groups enables more straightforward management of reaction conditions, which helps streamline the synthetic pathway.

(64) (C) As late as possible.

Protecting groups should be removed as late as possible in the synthetic sequence. This strategy also allows for maximal use of the protecting groups throughout the reaction sequence.

(65) (A) Boc (tert-butyloxycarbonyl).

The choice of protecting group depends on several factors. These include the stability of the group under various reaction conditions, its ease of introduction and removal, and its influence on the overall reactivity of the molecule. Commonly employed protecting groups include tert-butyloxycarbonyl (Boc) for amines and trimethylsilyl (TMS) for alcohols.

(66) (C) Nylon.

Synthetic polymers like nylon, polyester, and acrylic are staples in textiles. They are used in clothing, carpets, and industrial fabrics due to their strength, elasticity, and resistance to wear and tear. Nylon is found in hosiery and ropes, polyester in apparel and upholstery, and acrylic in sweaters and blankets. These fibers withstand abrasion and repeated washing. They can be engineered to possess specific properties, such as moisture-wicking or flame resistance.

(67) (B) Bulletproof vests.

The textile industry is expanding into technical and smart applications. Aramid polymers include Kevlar, which is used in bulletproof vests, and Gore-Tex (expanded PTFE), which is used to create waterproof, breathable fabrics. Conductive polymers, such as polyaniline, have been utilized to produce smart textiles for wearable electronics like heart rate-monitoring sensors. Sustainability is a growing focus, with recycled polyester derived from PET bottles and bio-based fibers, like lyocell, which help reduce environmental impact. However, microplastic pollution from synthetic textiles is a challenge. This concern has driven research into biodegradable fibers and washing machine filtration systems.

(68) (B) Ease of molding.

Polymers with ease of molding can be readily shaped into various forms when heated, making them highly suitable for applications like injection molding, extrusion, and blow molding in industrial manufacturing.

(69) (A) Polyamide.

With the rise of electric vehicles, polymers such as polyamide are utilized in battery housings due to their thermal stability. Self-healing polymeric coatings for scratch repair and recycled plastics for sustainability are also emerging. High-performance polymers must withstand extreme conditions. This issue has prompted research into advanced thermoplastics and composites. As autonomous vehicles develop, conductive polymers may play a role in sensors and lightweight structures.

(70) (C) Polycaprolactone (PCL).

In biomedical applications, polymers are transformative, which enable drug delivery systems, tissue engineering scaffolds, and medical implants. PLGA facilitates controlled drug release, while polycaprolactone (PCL) is used in 3D-printed implants for bone regeneration. Biocompatible and biodegradable polymers ensure safe interaction with tissues and harmless degradation in the body.

Test 6: Questions

(1) What is a common mistake in the application of IUPAC nomenclature?

(A) Select the shortest carbon chain as the parent chain.

(B) Assign the lowest locants to substituents.

(C) Include stereochemical descriptors.

(D) Alphabetize substituents correctly.

(2) When arranging substituents in alphabetical order for nomenclature purposes, which specific element should be disregarded?

(A) The root name of the substituent, which identifies the base structure.

(B) Multiplying prefixes such as di- or tri-, which indicate the number of identical groups.

(C) The locant numbers that indicate the positions of substituents on the parent chain.

(D) The priority of the functional group within the compound.

(3) What stereochemical descriptors are necessary for a compound that contains a chiral center?

(A) Cis/trans, which address geometric isomerism in certain structures.

(B) E/Z, which describe geometric isomers involving double bonds.

(C) R/S, which specify the absolute configuration of chiral centers.

(D) Ortho/meta, which are related to substitution patterns in aromatic compounds.

(4) In the context of IUPAC nomenclature, what is the correct suffix that is associated with an ester functional group?

(A) -oate.

(B) -amide.

(C) -al.

(D) -one.

(5) Which of the following compounds serves as an example of a common name that has been retained in IUPAC nomenclature?

(A) Hex-2-ene.

(B) Toluene.

(C) Butan-2-ol.

(D) Pentanoic acid.

(6) What makes naphthalene an aromatic compound?

(A) Its structure contains eight π electrons.

(B) It is non-planar.

(C) It possesses ten π electrons.

(D) It lacks conjugation.

(7) What type of reaction do aromatic compounds usually undergo?

(A) Addition.

(B) Elimination.

(C) Electrophilic substitution.

(D) Nucleophilic addition.

(8) What is a common mistake when applying Hückel's rule to determine aromaticity?

(A) Check for planarity of the molecular structure.

(B) Verify that the structure is cyclic.

(C) Count the number of π electrons present.

(D) Ignore the requirements for full conjugation within the system.

(9) Which bond is the shortest within the but-1-yne molecule?

(A) The C–H bond.

(B) The C–C single bond.

(C) The C≡C triple bond.

(D) The C–C double bond.

(10) How does conjugation affect the results observed in NMR spectroscopy?

(A) It shifts signals upfield.

(B) It reduces the intensity of signals observed in the spectrum.

(C) It eliminates proton signals from the spectrum entirely.

(D) It alters chemical shifts.

(11) All of the following are ways in which bases act as catalysts in chemical reactions except:

(A) They deprotonate a reactant.

(B) They generate a stronger nucleophile.

(C) They activate a leaving group.

(D) They protonate a reactant.

(12) Which of the following statements best explains why basicity tends to decrease within the same period of the periodic table as electronegativity increases?

(A) More electronegative atoms have larger atomic sizes, which reduces the availability of lone pairs.

(B) Electronegative atoms hold their lone pairs more tightly, which makes them less likely to donate them to a proton.

(C) Electronegative atoms have a higher tendency to form covalent bonds, which diminishes their basic character.

(D) More electronegative atoms have a higher number of protons, which causes reduced basicity.

(13) Which of the following statements best describes conformational isomers (or conformers)?

(A) Isomers that differ in the connectivity of atoms.

(B) Isomers that differ in spatial arrangements due to rotation around single bonds.

(C) Isomers that have the same molecular formula but different boiling points.

(D) Isomers that are isolated at room temperature.

(14) What does a Newman projection represent?

(A) The molecular formula of a compound.

(B) The spatial arrangement of atoms in a molecule.

(C) The conformation of a carbon-carbon single bond viewed along its axis.

(D) The steric interactions between different atoms.

(15) What does the x-axis of an energy diagram represent?

(A) Potential energy.

(B) Reaction coordinate.

(C) Reaction rate.

(D) Temperature.

(16) Which of the following indicates the energy difference between reactants and products in an energy diagram?

(A) Activation energy.

(B) Transition state.

(C) ΔG° or ΔH°.

(D) Reaction coordinate.

(17) Which of the following describes a transition state in a chemical reaction?

(A) A high-energy, unstable species.

(B) A stable product.

(C) An energy minimum.

(D) An intermediate.

(18) What is the rate-determining step in a multistep reaction?

(A) The fastest step.

(B) The step with the lowest activation energy.

(C) The step with the highest activation energy.

(D) The step involving the most reactants.

(19) Which of the following best describes a key feature of the E1 (unimolecular elimination) mechanism?

(A) The reaction occurs in a single concerted step.

(B) A strong base is always required.

(C) The rate-determining step involves carbocation formation.

(D) The reaction is stereospecific.

(20) At what temperature do E1 reactions generally occur?

(A) Lower.

(B) Moderate.

(C) Higher.

(D) Room temperature.

(21) What type of addition occurs when a nucleophile attacks the β-carbon of an α,β-unsaturated carbonyl compound?

(A) 1,2-addition.

(B) 1,3-addition.

(C) 1,4-addition.

(D) Elimination.

(22) What factor determines the regioselectivity of nucleophilic addition to α,β-unsaturated carbonyls?

(A) The temperature of the reaction.

(B) The nature of the solvent.

(C) The nature of the nucleophile and reaction conditions.

(D) The concentration of the carbonyl compound.

(23) Which of the following nucleophiles would likely favor a 1,2-addition to α,β-unsaturated carbonyl compounds?

(A) Alcohols.

(B) Enolates.

(C) Primary amines.

(D) Grignard reagents.

(24) What type of addition occurs when a nucleophile directly attacks the carbonyl carbon of an α,β-unsaturated carbonyl compound?

(A) 1,2-addition.

(B) 1,3-addition.

(C) 1,4-addition.

(D) Elimination.

(25) What is the first step in the nucleophilic addition to carbonyls?

(A) Protonation of the alkoxide.

(B) Nucleophilic attack.

(C) Formation of a double bond.

(D) Elimination of water.

(26) Which of the following products is formed when hydrogen cyanide (HCN) reacts with carbonyl compounds?

(A) Aldehyde.

(B) Alcohol.

(C) Cyanohydrin.

(D) Ketone.

(27) Which of the following products is formed when two equivalents of alcohol react with an aldehyde in the presence of an acid catalyst?

(A) Acetal.

(B) Hemiacetal.

(C) Ketone.

(D) Alcohol.

(28) Which of the following products is formed when a primary amine reacts with an aldehyde?

(A) Hemiacetal.

(B) Imine (Schiff base).

(C) Alcohol.

(D) Ketone.

(29) What is a molecular ion (M^+) in the context of mass spectroscopy?

(A) An ion formed by adding electrons.

(B) An ion formed by the removal of one electron from the parent molecule.

(C) A neutral atom.

(D) An ion formed by protonation.

(30) Which of the following is expected to happen during fragmentation in mass spectrometry?

(A) The molecular ion remains intact.

(B) The molecular ion breaks apart into smaller ions and neutral fragments.

(C) Only the base peak is formed.

(D) The sample is vaporized.

(31) Which of the following statements regarding the base peak in a mass spectrum is correct?

(A) The lowest peak in the spectrum.

(B) The peak corresponding to the molecular ion.

(C) The most abundant ion in the mass spectrum.

(D) An ion with the highest m/z value.

(32) What usually triggers fragmentation in mass spectrometry?

(A) Weak bonds.

(B) Strong bonds.

(C) High temperatures.

(D) Low pressures.

(33) What type of cleavage occurs next to heteroatoms or π systems?

(A) Homolytic cleavage.

(B) Heterolytic cleavage.

(C) α-Cleavage.

(D) β-Cleavage.

(34) Which of the following is formed during homolytic cleavage?

(A) A stable ion.

(B) A cation and an anion.

(C) Two radicals, one of which becomes an ion.

(D) A neutral molecule.

(35) What is the first step to perform in spectral interpretation?

(A) Propose a structure.

(B) Analyze NMR data.

(C) Determine the molecular formula.

(D) Verify the structure.

(36) What challenge is associated with the use of synthetic antioxidants?

(A) They are always ineffective.

(B) Their environmental persistence raises concerns.

(C) They are too inexpensive.

(D) They cannot be used in polymers.

(37) What is one practical application of studying radical-based enzyme mechanisms?

(A) Engineer microbes to produce biofuels or pharmaceuticals.

(B) Develop synthetic fertilizers.

(C) Create artificial intelligence.

(D) Enhance mineral extraction.

(38) What is a challenge in harnessing radical enzymes for industrial use?

(A) Their intricate mechanisms are difficult to understand.

(B) They are too simple.

(C) They are highly unstable.

(D) They require extreme conditions.

(39) Which of the following statements regarding the role of free radicals in disease is correct?

(A) They enhance cellular function.

(B) They promote healthy metabolism.

(C) They facilitate nutrient absorption.

(D) They damage DNA, proteins, and lipids.

(40) Which vitamin acts as an antioxidant within the human body?

(A) Vitamin A.

(B) Vitamin D.

(C) Vitamin C.

(D) Vitamin K.

(41) Which of the following statements regarding the challenge of using synthetic antioxidants is correct?

(A) They are too effective.

(B) They can disrupt beneficial radical signaling.

(C) They have no side effects.

(D) They are easily absorbed.

(42) Which of the following is an example of a strongly deactivating group in aromatic substitution reactions?

(A) Hydroxyl group (–OH).

(B) Nitro group (–NO_2).

(C) Alkyl group (–R).

(D) Amide (–$CONH_2$).

(43) What effect do electronegative atoms have on the aromatic ring?

(A) +M effect.

(B) –M effect.

(C) +I effect.

(D) –I effect.

(44) Which of the following groups is classified as a moderately deactivating group in aromatic substitution reactions?

(A) Carboxylic acid (–COOH).

(B) Alkyl group (–R).

(C) Trifluoromethyl group (–CF_3).

(D) Hydroxyl group (–OH).

(45) What is the main mechanism through which strongly deactivating groups withdraw electron density in aromatic substitution reactions?

(A) Inductive effects only.

(B) Resonance effects only.

(C) Both resonance and inductive effects.

(D) Hyperconjugation.

(46) What unique characteristic do halogens exhibit in terms of their effect on the aromatic ring?

(A) They only have a –I effect.

(B) They only have a +M effect.

(C) They have no effect on the ring.

(D) They exhibit both –I and +M effects.

(47) What distinguishes weakly deactivating groups from strongly deactivating groups?

(A) Weakly deactivating groups do not withdraw electrons.

(B) Weakly deactivating groups have a stronger inductive effect.

(C) Weakly deactivating groups are only found in alkyl chains.

(D) Weakly deactivating groups mainly use inductive effects, while strongly deactivating groups use both resonance and inductive effects.

(48) Which of the following products is formed when carboxylic acids react with amines?

(A) Esters.

(B) Carboxylate salts.

(C) Alcohols.

(D) Amides.

(49) Which reducing agent can convert carboxylic acids to alcohols?

(A) Sodium borohydride ($NaBH_4$).

(B) Lithium aluminum hydride ($LiAlH_4$).

(C) Hydrogen gas (H_2).

(D) Magnesium metal (Mg).

(50) What type of reaction occurs when a carboxylic acid reacts with an alcohol?

(A) Esterification.

(B) Neutralization.

(C) Decarboxylation.

(D) Reduction.

(51) Which reagent is commonly used to form acid halides from carboxylic acids?

(A) Sodium hydroxide (NaOH).

(B) Thionyl chloride ($SOCl_2$).

(C) Lithium aluminum hydride ($LiAlH_4$).

(D) Ethanol.

(52) Which statement regarding the reactivity of carboxylic acid derivatives is true?

(A) Anhydrides are more reactive than esters.

(B) Acid halides are less reactive than amides.

(C) Amides are the most reactive of all derivatives.

(D) Esters are more reactive than acid halides.

(53) Which of the following is formed by the reaction of a carboxylic acid with an alcohol?

(A) Amide.

(B) Acid halide.

(C) Anhydride.

(D) Ester.

(54) What is the overall outcome of the Michael reaction?

(A) Formation of a ketone.

(B) Formation of a new carbon-carbon bond.

(C) Formation of an alcohol.

(D) Elimination of water.

(55) All of the following factors affect the rate and regioselectivity of conjugate addition in the Michael reaction except:

(A) Nature of the nucleophile.

(B) Electrophilicity of the α,β-unsaturated carbonyl compound.

(C) Solvent used.

(D) The color of the reactants.

(56) Which strong base is commonly used to generate the enolate ion in the Robinson annulation?

(A) Sodium hydride.

(B) Sodium chloride.

(C) Potassium carbonate.

(D) Sodium bicarbonate.

(57) All of the following are benefits of the Robinson annulation except:

(A) Efficient construction of six-membered rings.

(B) Production of racemic mixtures.

(C) Controlled formation of specific stereoisomers.

(D) Ability to create multiple bonds and functional groups in a single reaction.

(58) Which method selectively reduces ketones to alcohols using a chiral oxazaborolidine catalyst?

(A) Corey-House synthesis.

(B) CBS reduction.

(C) Grignard reaction.

(D) Catalytic hydrogenation.

(59) What is the main purpose of the Sharpless epoxidation?

(A) Reduce ketones to alcohols.

(B) Oxidize alcohols to carboxylic acids.

(C) Couple alkyl halides.

(D) Epoxidize allylic alcohols in a stereoselective manner.

(60) Which of the following describes chemoselectivity?

(A) The ability of a reagent to select specific solvents.

(B) The ability of a reagent to react preferentially with one functional group over another.

(C) The ability to form multiple products in a reaction.

(D) The speed at which a reaction occurs.

(61) What is the benefit of delaying the removal of protecting groups?

(A) Enhance the reactivity of the molecule.

(B) Simplify the synthetic pathway.

(C) Speed up the overall synthesis.

(D) Avoid unwanted reactions from the early exposure of functional groups.

(62) What happens if reactive functional groups are left unprotected during synthesis?

(A) The synthesis will be faster.

(B) The final product will be simpler.

(C) The overall yield will increase.

(D) Unwanted side reactions may occur.

(63) Which of the following factors should be given priority in terms of practicality in a multistep synthetic plan?

(A) The aesthetic appeal of the final product.

(B) The cost of reagents and availability of starting materials.

(C) The length of the synthesis.

(D) The complexity of the reactions.

(64) Which strategy employs chiral catalysts to promote the formation of chiral products in high enantiomeric excess?

(A) Convergent synthesis.

(B) Domino reactions.

(C) Chiral pool synthesis.

(D) Asymmetric catalysis.

(65) Which synthesis method combines two or more complex fragments that are synthesized separately before being joined in a final step?

(A) Convergent synthesis.

(B) Domino reactions.

(C) Chiral pool synthesis.

(D) Asymmetric catalysis.

(66) Which of the following is considered a safe and non-hazardous solvent?

(A) Chloroform.

(B) Benzene.

(C) Supercritical carbon dioxide ($scCO_2$).

(D) Methanol.

(67) Which chemical reagent is commonly used in presumptive testing to detect the presence of amphetamines?

(A) Marquis Reagent.

(B) Benedict's Solution.

(C) Tollen's Reagent.

(D) Lucas Reagent.

(68) What color indicates the presence of cocaine in the Scott Test?

(A) Purple.

(B) Orange-brown.

(C) Blue.

(D) Green.

(69) What test is considered the gold standard for drug identification?

(A) Scott Test.

(B) Duquenois-Levine Test.

(C) High-performance liquid chromatography.

(D) Gas chromatography-mass spectrometry (GC-MS).

(70) Which of the following statements regarding the mechanism of action of the luminol test is correct?

(A) It changes color with pH.

(B) It detects enzymes in saliva.

(C) It identifies bacterial presence.

(D) It reacts with hemoglobin to emit blue light.

Test 6: Answers and Explanations

(1) (A) Select the shortest carbon chain as the parent chain.

A frequent error in IUPAC nomenclature is choosing the shortest carbon chain as the parent chain instead of the longest continuous carbon chain that includes the functional groups and substituents. The longest chain is used to accurately identify the compound's structure and ensure that all relevant functional groups are included in the naming process.

(2) (B) Multiplying prefixes such as di- or tri-, which indicate the number of identical groups.

When chemists alphabetize substituents in the naming process, they must ignore any multiplying prefixes like di- or tri-. Only the root name is used to determine the order (for instance, dimethyl would be alphabetized under /m/). In contrast, the root name, locant numbers, and functional group priority are all relevant elements that should not be disregarded during this process.

(3) (C) R/S, which specify the absolute configuration of chiral centers.

For compounds that possess a chiral center, it is vital to use the R/S descriptors to indicate their stereochemistry accurately. The cis/trans and E/Z designations apply to geometric isomers, particularly those involving double bonds. The ortho/meta terminology pertains to the arrangement of substituents on aromatic rings.

(4) (A) -oate.

In IUPAC nomenclature, esters are named by using the suffix -oate. This suffix indicates the presence of the ester functional group, which is derived from a carboxylic acid. For example, methyl acetate is an ester formed from acetic acid and methanol, where the term acetate reflects the ester functionality.

(5) (B) Toluene.

Toluene is recognized as a common name for the compound methylbenzene, and it has been retained in official IUPAC nomenclature. Hex-2-ene, butan-2-ol, and pentanoic acid are examples of systematic names rather than common names.

(6) (C) It possesses ten π electrons.

Naphthalene has a total of ten π electrons (fitting the $4n + 2$ rule with $n=2$) and is cyclic, planar, and fully conjugated, which qualifies it as an aromatic compound. Eight π electrons would classify a compound as anti-aromatic.

(7) (C) Electrophilic substitution.

Aromatic compounds are stabilized by resonance. They predominantly undergo electrophilic substitution reactions to maintain their π electron system. Addition reactions are for reactions involving alkenes, and nucleophilic addition reactions are specific to carbonyl compounds, not aromatic systems.

(8) (D) Ignore the requirements for full conjugation within the system.

A significant mistake when applying Hückel's rule is neglecting to ensure that the molecule has complete conjugation. This oversight can result in the misclassification of non-aromatic compounds as aromatic.

(9) (C) The C≡C triple bond.

The C≡C triple bond in but-1-yne is the shortest bond due to its high bond order, which comprises three bonds between the carbon atoms. The C–H and C–C single bonds are longer. Bond length is as follows: triple bonds < double bonds < single bonds.

(10) (D) It alters chemical shifts.

Conjugation changes the electronic environment around the nuclei, which alters the chemical shifts observed in NMR spectra. Conjugation often causes downfield shifts, and proton signals remain present.

(11) (D) They protonate a reactant.

Bases can catalyze chemical reactions by deprotonating a reactant to generate a stronger nucleophile or by activating a leaving group. For example, in aldol condensation under basic conditions, a base (e.g., OH^-) deprotonates an α-hydrogen of a carbonyl compound, which creates an enolate ion, a strong nucleophile.

(12) (B) Electronegative atoms hold their lone pairs more tightly, which makes them less likely to donate them to a proton.

Basicity tends to decrease within the same period as electronegativity increases. A more electronegative atom holds its lone pair more tightly and makes it less likely to donate it to a proton.

(13) (B) Isomers that differ in spatial arrangements due to rotation around single bonds.

A conformational isomer (or conformer) has a different spatial arrangement of atoms due to rotation around single bonds. It interconverts rapidly at room temperature and is often not isolable. Conformational analysis studies the relative energies and populations of different conformers.

(14) (C) The conformation of a carbon-carbon single bond viewed along its axis.

A Newman projection visually represents the conformation of a carbon-carbon single bond by looking directly along the bond's axis. The carbon atom closer to the viewer is depicted as a dot with three bonds radiating from it. The further carbon appears as a circle, with its three attached bonds extending from its edge.

(15) (B) Reaction coordinate.

An energy (reaction coordinate) diagram illustrates the energy changes that occur during a chemical reaction, which depicts the progression from reactants to products. The x-axis represents the reaction coordinate and indicates the reaction's progress, while the y-axis shows the potential energy of the system.

(16) (C) $\Delta G°$ or $\Delta H°$.

The energy difference between reactants and products ($\Delta G°$ or $\Delta H°$) determines whether the reaction is exergonic (releases energy, therefore products are lower in energy) or endergonic (requires energy, therefore products are higher in energy).

(17) (A) A high-energy, unstable species.

A transition state is a high-energy, unstable species that represents the point of maximum energy along the reaction pathway. It corresponds to an energy maximum on the diagram and indicates where bonds are breaking and forming.

(18) (C) The step with the highest activation energy.

In a multistep reaction, the rate-determining (-limiting) step has the highest activation energy and is the slowest step in the reaction mechanism. This step limits the overall speed of the reaction and thus dictates the rate at which the reaction progresses.

(19) (C) The rate-determining step involves carbocation formation.

In the E1 mechanism, the first and slowest step is the loss of the leaving group, forming a carbocation intermediate. This unimolecular step determines the rate of the reaction and is independent of the base concentration.

(20) (C) Higher.

E1 reactions are favored at higher temperatures since they involve the formation of a carbocation intermediate, which can be energetically costly. Increasing the temperature helps to overcome this barrier and favors the formation of products through elimination.

(21) (C) 1,4-addition.

In 1,4-addition (conjugate addition or Michael addition), the nucleophile attacks the β-carbon of the α,β-unsaturated system. This occurs because the π electrons of the double bond and the carbonyl group are conjugated, and the β-carbon exhibits a partial positive charge due to resonance.

(22) (C) The nature of the nucleophile and reaction conditions.

The regioselectivity of nucleophilic addition to α,β-unsaturated carbonyls depends on the nature of the nucleophile and the reaction conditions. Strong, hard nucleophiles (e.g., Grignard, organolithium, and hydride reagents) tend to favor 1,2-addition, while softer, more stabilized nucleophiles (e.g., enolates, cyanide, thiols, and amines) usually favor 1,4-addition.

(23) (D) Grignard reagents.

Grignard reagents are strong carbon nucleophiles that usually favor 1,2-addition due to their high reactivity and ability to attack the electrophilic carbon of the carbonyl directly.

(24) (A) 1,2-addition.

Carbonyl compounds with a double bond adjacent to the carbonyl group (α,β-unsaturated carbonyls) can undergo two types of nucleophilic addition. In 1,2-addition (direct addition), the nucleophile directly attacks the carbonyl carbon.

(25) (B) Nucleophilic attack.

In a nucleophilic attack, a nucleophile attacks the electrophilic carbonyl carbon. The π bond between carbon and oxygen breaks. This reaction transfers electrons to the oxygen atom and creates an alkoxide intermediate with a negative charge on the oxygen.

(26) (C) Cyanohydrin.

Hydrogen cyanide (HCN) adds to aldehydes and ketones to form cyanohydrins, which are useful intermediates in organic synthesis. The cyanide ion (CN^-) acts as the nucleophile.

(27) (A) Acetal.

Alcohols (in the presence of acid catalyst) can add to aldehydes to form hemiacetals and acetals (if two equivalents of alcohol are used). Ketones form hemiketals and ketals, which serve as important protecting groups for aldehydes and ketones.

(28) (B) Imine (Schiff base).

Primary amines (RNH_2) react with aldehydes and ketones to produce imines (Schiff bases) with the elimination of water, while secondary amines (R_2NH) yield enamines.

(29) (B) An ion formed by the removal of one electron from the parent molecule.

A molecular ion (M^+) is formed by the removal of one electron from the parent molecule. Its m/z value corresponds to the compound's molecular weight. The molecular ion peak may not always be prominent or even observed, especially for unstable molecules.

(30) (B) The molecular ion breaks apart into smaller ions and neutral fragments.

During fragmentation, the molecular ion is often unstable and can break apart into smaller ions and neutral fragments. The resulting pattern of fragment ions is characteristic of the molecule's structure and can be used to deduce structural features.

(31) (C) The most abundant ion in the mass spectrum.

The most abundant ion in the mass spectrum is known as the base peak. It is represented by the tallest peak in the spectrum. This peak is significant because its relative abundance is defined as 100%.

(32) (A) Weak bonds.

Fragmentation occurs at weak bonds during mass spectrometry. It causes the formation of ions that can be analyzed, which helps identify functional groups and molecular weight.

(33) (C) α-Cleavage.

The fragmentation pathway, α-cleavage, occurs adjacent to heteroatoms or π systems. It results in the formation of resonance-stabilized ions. These ions are more stable and detectable in the mass spectrum, and they provide valuable information about the molecular structure.

(34) (C) Two radicals, one of which becomes an ion.

Homolytic cleavage results in the formation of two radicals when a bond breaks symmetrically. One of these radicals can become an ion, which allows for its measurement in mass spectrometry.

(35) (C) Determine the molecular formula.

The first step in spectral interpretation involves determining the molecular formula. It provides essential information about the compound's composition and helps calculate the degree of unsaturation.

(36) (B) Their environmental persistence raises concerns.

The use of antioxidants enhances product durability and safety. Synthetic antioxidants are highly effective. However, concerns about their environmental persistence have spurred research into bio-based alternatives, such as tocopherols (vitamin E) or plant-derived polyphenols.

(37) (A) Engineer microbes to produce biofuels or pharmaceuticals.

Radical-based enzyme mechanisms are highly efficient and selective, which inspires the development of synthetic catalysts. Their study has led to applications in biotechnology, such as engineering microbes to produce biofuels or pharmaceuticals.

(38) (A) Their intricate mechanisms are difficult to understand.

Challenges of utilizing radical-based enzymes include understanding their intricate mechanisms and harnessing them for industrial use. Advances in structural biology, such as cryo-electron microscopy, continue to elucidate these mechanisms, while synthetic biology works to expand its applications in sustainable chemical production.

(39) (D) They damage DNA, proteins, and lipids.

Free radicals, such as superoxide anion ($O_2^{\bullet -}$), hydroxyl radical (•OH), and lipid peroxyl radicals (ROO•), are generated during metabolic processes or through exposure to radiation, pollutants, or toxins. These highly reactive species can damage DNA, proteins, and lipids, which contribute to the aging process and diseases such as cancer, cardiovascular disorders, and neurodegenerative conditions. For example, lipid peroxidation in cell membranes, driven by ROO•, disrupts cellular integrity, which accelerates disease progression.

(40) (C) Vitamin C.

Biological systems counteract radical damage through sophisticated antioxidant defenses, such as enzymes like superoxide dismutase (SOD), catalase, and glutathione peroxidase, and small-molecule antioxidants like vitamins C and E and glutathione.

(41) (B) They can disrupt beneficial radical signaling.

Antioxidant defenses are used to prevent oxidative stress-related diseases. For instance, vitamin E scavenges peroxyl radicals to protect lipid membranes. Research is exploring synthetic antioxidants inspired by natural systems, such as mimetics of SOD, for therapeutic use. Challenges include delivering antioxidants to specific cellular sites and avoiding disruption of beneficial radical signaling.

(42) (B) Nitro group ($-NO_2$).

Strongly deactivating groups withdraw electron density from the ring through resonance and strong inductive effects. Examples include nitro group ($-NO_2$), cyano group (–CN), trifluoromethyl group ($-CF_3$), and quaternary ammonium groups ($-NR_3^+$).

(43) (D) –I effect.

Electronegative atoms or groups withdraw electron density through inductive effects (–I effect). Deactivating groups with π systems directly attached to the ring can withdraw electron density through resonance (–M or –R effects), which decreases electron density, especially at the ortho and para positions.

(44) (A) Carboxylic acid (–COOH).

Moderately deactivating groups withdraw electron density through resonance involving a carbonyl group directly attached to the ring. The electronegative oxygen atom pulls electron density away from the ring. Examples include carboxylic acids (–COOH), esters (–COOR), amides ($–CONH_2$), aldehydes (–CHO), and ketones (–COR).

(45) (C) Both resonance and inductive effects.

Deactivating groups are substituents on an aromatic ring that decrease the electron density of the ring. This makes the ring less nucleophilic and less reactive toward electrophiles compared to benzene. These groups strongly withdraw electron density from the ring through resonance and strong inductive effects.

(46) (D) They exhibit both –I and +M effects.

Halogens (–F, –Cl, –Br, –I) are unique. They are deactivating due to their electronegativity (–I effect), which withdraws electron density from the ring. However, they also possess lone pairs that can participate in resonance (+M effect), although this effect is weaker than their inductive withdrawal.

(47) (D) Weakly deactivating groups mainly use inductive effects, while strongly deactivating groups use both resonance and inductive effects.

Strongly deactivating groups withdraw electron density through both resonance and strong inductive effects. This significantly reduces aromatic reactivity. Weakly deactivating groups mainly use inductive effects with weaker resonance contributions.

(48) (D) Amides.

Carboxylic acids react with amines to form amides. This reaction usually requires the activation of the carboxylic acid. For example, converting it to an acid chloride or an activated ester.

(49) (B) Lithium aluminum hydride ($LiAlH_4$).

Carboxylic acids can be reduced to primary alcohols using strong reducing agents such as lithium aluminum hydride ($LiAlH_4$). They can also undergo decarboxylation under certain conditions.

(50) (A) Esterification.

Carboxylic acids react with alcohols in the presence of an acid catalyst to form esters. This reaction is known as Fischer esterification. Carboxylic acids also react with bases to form carboxylate salts through neutralization.

(51) (B) Thionyl chloride ($SOCl_2$).

Acid halides (RCOX, where X = Cl, Br) are formed by the reaction of a carboxylic acid with a halogenating agent such as thionyl chloride ($SOCl_2$) or phosphorus pentachloride (PCl_5).

(52) (A) Anhydrides are more reactive than esters.

The reactivity of carboxylic acid derivatives toward nucleophilic acyl substitution depends on the nature of the leaving group. Acid halides are the most reactive, followed by anhydrides, esters, and amides.

(53) (D) Ester.

When a carboxylic acid reacts with an alcohol, the reaction is called esterification, which results in the formation of an ester. This process usually requires an acid catalyst and involves the elimination of water.

(54) (B) Formation of a new carbon-carbon bond.

The Michael reaction is the conjugate addition of an enolate to an α,β-unsaturated carbonyl compound. This reaction is a powerful method for forming carbon-carbon bonds and is widely used in organic synthesis.

(55) (D) The color of the reactants.

Several factors can influence the rate and regioselectivity of conjugate addition reactions.

- Nature of the nucleophile: Stronger nucleophiles tend to react faster.
- Nature of the α,β-unsaturated carbonyl compound: EWGs on the carbonyl compound increase its electrophilicity and make it more reactive.
- Steric effects: Steric hindrance can affect the regioselectivity of the reaction. Bulky nucleophiles may prefer to attack the less hindered β-carbon.
- Solvent: The solvent can also influence the rate and regioselectivity of the reaction.

(56) (A) Sodium hydride.

The first phase of the Robinson annulation involves a Michael addition, where a nucleophilic attack occurs. This process begins with the formation of an enolate ion, generated when a strong base (such as sodium hydride or potassium tert-butoxide) deprotonates a carbonyl compound at its α-position

(57) (B) Production of racemic mixtures.

The Robinson annulation is renowned for the following reasons.

- Versatility: This reaction efficiently constructs six-membered rings, a prevalent motif in organic molecules. Its ability to generate complex cyclic structures is invaluable in synthetic organic chemistry.
- Stereoselectivity: The reaction often produces specific stereoisomers, which are critical in the synthesis of biologically active compounds. The controlled formation of stereocenters significantly enhances the utility of the Robinson annulation in pharmaceutical applications.
- Synthetic utility: The Robinson annulation serves as a key tool for building intricate structures, particularly those found in natural products. Its capability to create multiple bonds and functional groups in a single reaction sequence streamlines synthetic pathways.

(58) (B) CBS reduction.

The Corey-Itsuno reduction is also known as the Corey-Bakshi-Shibata (CBS) reduction. This method is used to selectively and enantioselectively reduce ketones to chiral alcohols. It utilizes a chiral oxazaborolidine catalyst and borane as a reducing agent. This reaction synthesizes chiral secondary alcohols, which are important building blocks in the synthesis of various natural products and pharmaceuticals.

(59) (D) Epoxidize allylic alcohols in a stereoselective manner.

The Sharpless epoxidation is a stereoselective reaction that converts allylic alcohols into epoxides using a titanium catalyst and chiral tartrate esters. It allows for the selective synthesis of cyclic ethers.

(60) (B) The ability of a reagent to react preferentially with one functional group over another.

Chemoselectivity is the preferential reaction of a chemical reagent with one specific functional group in a molecule, even when other functional groups are present and potentially reactive.

(61) (D) Avoid unwanted reactions from the early exposure of functional groups.

Protecting groups should be removed as late as possible in the synthetic sequence. Delaying the removal of protecting groups helps avoid unwanted reactions that could occur if reactive functional groups are exposed too early. By keeping these groups intact until the final stages of synthesis, chemists can maintain the integrity of the desired functional groups and ensure that the final product reflects the intended molecular architecture. This strategy also allows for maximal use of the protecting groups throughout the reaction sequence.

(62) (D) Unwanted side reactions may occur.

When reactive functional groups, such as alcohols, amines, or carboxylic acids, are left unprotected during chemical synthesis, they can participate in side reactions. These side reactions result in undesired products. This complicates the synthetic pathway and lowers the overall yield of the desired product. Protecting these functional groups maintains control over reactions and ensures successful transformations.

(63) (B) The cost of reagents and availability of starting materials.

Designing a multistep synthesis is a complex and challenging task that requires a deep understanding of organic reactions, functional group chemistry, and strategic planning. Chemists must consider the practicality of each step, such as the cost of reagents, the availability of starting materials, and the potential for side reactions.

(64) (D) Asymmetric catalysis.

Asymmetric catalysis uses chiral catalysts to promote the formation of chiral products in high enantiomeric excess. This strategy is widely used in the synthesis of pharmaceuticals and other fine chemicals. Chiral pool synthesis uses readily available chiral starting materials (e.g., amino acids, carbohydrates, terpenes) to synthesize chiral target molecules. This strategy can be beneficial for synthesizing natural products.

(65) (A) Convergent synthesis.

In a convergent synthesis, two or more complex fragments are synthesized separately and then joined together in a final step. This strategy can be more efficient than a linear synthesis, in which the molecule is built up step-by-step from one end. Domino reactions (i.e., cascade or tandem reactions) are sequences of two or more reactions that occur in a single pot without the isolation of intermediates. They can significantly shorten a synthesis and improve its efficiency.

(66) (C) Supercritical carbon dioxide ($scCO_2$).

One of the cornerstones of green chemistry is the replacement of hazardous organic solvents, such as chloroform or benzene, with safer alternatives like water, supercritical carbon dioxide ($scCO_2$), and ionic liquids. Supercritical CO_2, which exists above its critical temperature and pressure, provides tunable properties, is recyclable, and minimizes waste. It is used in processes like caffeine extraction from coffee beans and leaves no toxic residues.

(67) (A) Marquis Reagent.

The Marquis Reagent is commonly used in presumptive drug testing. When it reacts with amphetamines, it typically produces an orange-brown color, indicating the possible presence of this class of stimulants.

(68) (C) Blue.

The Scott Test is specifically used for detecting cocaine. It employs cobalt thiocyanate, which forms a blue complex in the presence of cocaine. The simplicity and speed of this test allow law enforcement to make quick decisions regarding potential drug-related activities.

(69) (D) Gas chromatography-mass spectrometry (GC-MS).

Gas chromatography-mass spectrometry (GC-MS) is a technique that separates vaporized compounds based on their volatility and interactions with the chromatographic column. After separation, the compounds are fragmented to produce a unique mass spectrum, which can be compared against reference libraries. GC-MS is considered the gold standard for drug identification due to its high sensitivity and specificity.

(70) (D) It reacts with hemoglobin to emit blue light.

Organic chemistry underpins various tests that identify biological fluids at crime scenes and provides critical evidence in sexual assault cases and homicides. Luminol emits a blue chemiluminescence when it reacts with iron in hemoglobin and reveals trace bloodstains that may not be visible to the naked eye. This test is particularly useful for detecting blood at crime scenes where cleanup efforts may have been made.

Access the digital flash cards!

Scan the QR code below with your phone and you will be given a link to Google Drive where you can download the files.

If the QR code does not work for you, please contact us at info@newstonetestprep.com and mention the title of this book.

Made in United States
Orlando, FL
02 December 2025